Statistics
for
Social Change

Lucy Horwitz & Lou Ferleger

SOUTH END PRESS   BOSTON

96121

Library of Congress Card Number: 80-52136

ISBN 0-89608-033-1
ISBN 0-89608-034-X

Cover design and interior art work by Michael Prendergast.
Typesetting by the workers' cooperative at The Colonial Cooperative Press in Clinton, Massachusetts.
Design and paste up by the South End Press collective.

TABLE OF CONTENTS

To Our Students
For and With Whom
This Book Was Written

PREFACE

Statistics has in recent years become a subject of such importance in our daily lives that it can no longer be ignored by anyone who wants to understand the world we live in. Questions like "Does living near a nuclear power plant increase one's chance of having cancer?" "Are we headed for another recession?" or "Will the universe go on expanding forever?" can only be dealt with intelligently through an understanding of the basic statistical concepts which are used equally in attempts to find and to obscure the answers.

Statistics originally developed as a political tool of the eugenics enthusiasts. They have been used politically ever since by advocates of every possible position. At present, statistics are used to convince us of any number of politically expedient lies: the oil companies are losing money, the unemployment problem is being solved, inflation is being brought under control, and so on. But statistics can also be used to ferret out the truth.

Most people, bombarded daily by realms of statistical data, have conflicting reactions. They feel that all mathematical data is beyond their comprehension, but being numerical, must contain some truth. At the same time, there is a deep feeling of distrust. Numbers don't lie but statisticians do. As long as the data themselves are incomprehensible, these conflicting feelings cannot be sorted out or resolved. The result is confusion. The main goal of this text, then, is to dispel that confusion by helping the reader to sort out the misconceptions, misuses and outright lies from the useful applications; to replace false confidence as well as misplaced skepticism with a clear but critical understanding of statistics as a tool for viewing our society.

This text attempts to do that by concentrating on the concepts and applications of elementary statistics rather than on the mathematical derivation of formulas or on calculation procedures. Each chapter is divided into two parts, with Part A explaining the basic concepts and Part B giving examples and explanations of applications and common misapplications of those concepts.

The topics covered in this text include the usual selection for an introduction to descriptive and inferential statistics: frequency distributions; measures of central tendency and variability; probability and normal distributions; hypothesis testing and estimation; regression and correlation.

But this book also includes topics often neglected in other texts, beginning with an introductory chapter on deductive and inductive logic which provides a context for the study of statistics. There are also chapters on economic indicators and indices and on sampling and polling. These topics are of particular importance for those interested in applying statistics to the real world, and it is hoped that teachers and students will find time to cover all the material included. However, it is quite possible to omit these chapters and still have a coherent text. The two review chapters on graphs and percents may also be omitted for students with a strong background in these areas.

The text can be used in a number of different ways. It was developed in a one-semester introductory statistics course and could be used in a variety of courses of this kind with shifts in emphasis for students with different interests and degrees of mathematical sophistication. (Very little actual mathematical knowledge is required or assumed. One year of high school algebra should be sufficient.) This book can also be used as a supplemental text in various social science courses. In addition, it is intended as a useful resource for those who wish to learn statistics on their own. However it is used, the results will be best if the student works all the exercises, particularly the ones within the chapters. Statistics, like any other subject, cannot be learned passively. The more active a part the student takes in working through the material, applying it to daily experience and readings, the more thoroughly the contents will be mastered.

Many people have read the whole manuscript or parts of it and provided us with valuable comments and suggestions. We would like to express our gratitude to S. Schiavo-Campo, Jim Campen, Marilyn Frankenstein, Bill Fried, Marlene Fried, Roger Gottlieb, Chris Gulick, Barbara Herman, Carol Ivan, Paul O'Keefe, Michael Prendergast, Pat Walker and Paula Voos. We thank Arthur MacEwan for suggesting the title. Our special gratitude to Dick Cluster, our editor, whose many excellent suggestions led to substantial improvements in both style and content. Our students, especially those at CPCS, played a vital role in the development of the manuscript. And finally one of us (Horwitz) acknowledges her debt to Diane Hopkins, whose help and encouragement in the early stages got this project off the ground, and whose continued encouragement and technical assistance, along with that of Frederick Hamre, helped bring it to its conclusion.*†

* We are also grateful to the Literary Executor of the late Sir Ronald A. Fischer, F.R.S., to Dr. Frank Yates, F.R.S. and to Longman Group Ltd., London for permission to reprint Tables from their book *Statistical Tables for Biological, Agricultural and Medical Research.* (6th edition, 1974.)
† Solutions Manual is available upon request from the publisher.

INTRODUCTION

What Is Statistics?

The word statistics has two quite separate meanings in our language. First it means a collection of numbers which are considered, for some reason or other, interesting or important. Examples which come to mind include anything from the results of the latest government census to the batting averages reported in the sports pages. In this sense a **statistic** is a numerical fact. When you read in the paper that inflation last year was at the rate of 12%, you recognize a statistic.

The second meaning of the word **statistics** is the branch of applied mathematics devoted to the task of collecting, organizing and interpreting the numerical data we were talking about in describing the first sense of the word. For historical reasons, we use the same word for both the numbers and the science that deals with them. Originally, a statistician was simply someone who collected the statistics, or numerical facts, that the government needed in order to operate. Gradually techniques were developed, first to make it clearer what the numbers meant, and then to use these data to reach conclusions and make predictions, not only in government, but in industry and in most branches of the physical, biological and social sciences.

Modern statistics is divided into two branches: descriptive and inferential statistics. **Descriptive statistics** provides the tools for organizing and analyzing numerical data. It does just what its name implies: helps us to describe things numerically. Suppose we have a large set of numbers which measure something, such as test scores. Just looking at all these numbers would be confusing; we need some way to organize them so that they make sense. We can group the numbers in a variety of ways, we can tabulate them, or we can display them pictorially in graphs. We can also make sense of the data by calculating certain numbers which tell us about the whole set; a familiar example would be an average. Instead of looking at 200 test scores, we can average the scores (add them all up and divide by 200) to get an approximate sense of how well the students did. Or we may want to see how spread out the scores were. Then we would calculate numbers like the range (the difference between the highest and lowest scores) or the standard deviation (which we will learn to calculate later).

So descriptive statistics gives us ways of reducing a large set of numbers to a few numbers that describe the whole set. We may use descriptive statistics when we want to know the average income in a city, or how much unemployment there is among different groups, or how well our children are learning to read. The tools of descriptive statistics also help us to solve problems in inferential statistics.

Inferential statistics goes beyond description to draw conclusions or make predictions. That is, statistical techniques are used to make decisions about which interpretation of the data is more likely to reflect reality, allowing us to use the data to support arguments and hopefully to extend our knowledge of the subject. We might use inferential statistics to predict whether incomes are likely to increase or decrease next year, to decide what are the most important factors causing unemployment, or to see whether our children's reading skills are better or worse than others across the country.

The most important thing about inferential statistics is that it is a method for dealing with uncertainty by helping us make decisions when not enough information is known. Sometimes the uncertainty is about quantity. While once it may have been perfectly feasible for the King to send his "statistician" out to count all the sheep in his domain, it's hard to imagine the President of the U.S. doing the same thing. And yet the U.S. Government, no doubt publishes information on sheep herds, as well as many more, harder to get pieces of information. Does someone go to each household in the U.S. to see how many people are unemployed? Is the output of each mine, mill and factory tabulated to find the Gross National Product? Are the prices in every store in the country checked to calculate the Consumer Price Index? Of course not. Our world has become too large and complex to allow, in most cases, the certainty of counting. Just as we estimate when we are faced with a crowd too large to count, many government statistics are also the result of estimation. But while looking at a crowd and pulling a number out of the air gives us only the roughest kind of estimate, statistics provides us with more useful methods.

Estimation procedures involve such techniques as selecting a **sample,** or small group which will accurately represent the whole group. If we are interested in estimating something which cannot be directly measured, such as people's opinions, we must devise questionnaires or other tools for collecting the data. Then we use descriptive techniques to describe the data we have collected, and finally we must calculate the likelihood and size of error in our estimate. So inferential statistics offers techniques for reducing uncertainty, not only by making our estimates more accurate, but also by giving us information about possible errors.

Sometimes the uncertainty involves questions beyond mere numbers. We may be confronted with uncertainty about causes and effects, about trends, or about comparisons. For example, we may notice that executives often develop ulcers, so we may assume that ulcers are caused by the kind of stress executives are subject to. But this interpretation is only one possibility among many others. It may be that there is a certain type of person who has qualities which make him a good executive and which also make him subject to ulcers. Everyone may suffer equally from ulcers but executives, who get better medical attention, may have their ulcers diagnosed more often. Or it may be that a combination of all these factors is at work. While statistics alone cannot settle a question of cause and effect, careful experimental design can shed much light on causal relationships. Each of the possibilities to be tested is a hypothesis. In statistics we use the word **hypothesis** to mean any statement which we are trying to prove or disprove. And inferential statistics provides us with mathematical tools for testing these hypotheses to see which are more likely to be real explanations. These mathematical tools will never give us certain answers; they will only point to the more probable ones and thus help us make more informed decisions in the face of unavoidable uncertainty.

Statistical Accuracy

We have said that statistics is a method for dealing with uncertainty but also that the answers it provides are themselves uncertain. How is this possible when statistics is a branch of mathematics, which is supposed to be the most exact of all sciences? To understand this apparent paradox, it may help to think for a moment about what is meant by calling mathematics the exact science.

If I add 2 and 2, I will get 4, today, tomorrow or any other time. If you do it, you will get the same answer and so will anyone else who understands the concepts of number and addition. This idea can be extended to the most complicated calculation. Once the conditions of a mathematical exercise are stated, there is one and only one correct answer, which can be checked and agreed to by everyone with the knowledge necessary to understand that particular calculation. But what happens when we apply this most exact science to an actual life situation? I may measure two pieces of meat on my scale and get 2 lbs. + 2 lbs. = 4 lbs. You may weigh the same pieces on the same scale and get 2.1 lbs. + 1.95 lbs. = 4.05 lbs. If we had a very accurate scale, we could continue to weigh the meat over and over again and get slightly different answers depending on where it was placed on the scale, the angle from which the dial was read, what the air currents in the room were doing. And finally, if we continued long enough, the weights would decline as moisture evaporated from the meat. There could never be a "right" answer; yet in each case, after we finished the weighing, we could agree that our additions were correct.

The case of statistics is very much the same. Descriptive statistics provides us with formulas for calculating quantities which help us to describe a set of numbers. One example we have already mentioned is the average. Suppose we weigh our two pieces of meat a hundred times. We find that the average weight of those hundred weighings is 4.05 lbs. Anyone could check our arithmetic to see that this was indeed the average. But is this the best answer to what is the "accurate" weight? Perhaps a better answer would be to use the mode, the weight we recorded most often. If out of 100 weighings, the weight 4 lbs. was the most frequent, perhaps we should use that as the most "accurate" weight. The calculations are exact, but their applications are open to different interpretations.

Or take an example from inferential statistics. We want to investigate the connection between pornography and rape. We count the number of pornographic movies shown in a given city over a period of ten years and we count the number of rapes reported during the same period. Then we use a statistical formula for figuring what is called the correlation coefficient. We get a numerical answer. Anyone who agrees with us that we have accurately counted the number of pornographic movies shown and the number of rapes reported will be able to check and get the same result for the correlation coefficient. But again, what does it mean? I may say that the number I get proves that showing pornographic movies causes an increase in the number of rapes. My critics, although they do the same computation and get the same numerical answer may object to my conclusion on one or more of the following grounds:

1. There are not really more rapes, it is just that more are being reported.
2. The increase in both pornographic movies shown and rapes reported are due to changing sexual mores.
3. Both increases are due to socio-economic factors such as the frustrations and alienation of modern life.

When scientists use more sophisticated statistical tests to check their hypotheses, they encounter not only problems of interpretation, but also questions of which of a series of tests most accurately applies to a particular problem. Then, once a test has been selected and applied, the answer to the question of whether or not to accept the hypotheses will not be a "yes" or "no" answer. Rather, the mathematical calculation will yield a number which indicates the probability that the hypothesis is correct. Suppose we test the hypothesis that there is no connection between the numbers of pornographic movies shown and the number of rapes reported. When we do our calculations, using statistical formulas which anyone may duplicate, we will get a result which tells us, for example, that the odds are 100 to 1 that our hypothesis is incorrect. Or it may tell us that there is a 50-50 chance of being right or being wrong.

Statistics, then, is an exact science in that it provides us with a number of useful tools, in the form of mathematical formulas and procedures, which are exact in that they can always be duplicated. But these formulas and procedures are of interest only when we apply them to real life situations, and then we must again confront uncertainties.

Why Study Statistics?

Whatever the quality of the answers statistics provides, there can be no question about the quantity of statistical data that we are subjected to or the huge impact statistics has on all our lives. It is impossible to open a newspaper without being confronted with examples of statistics, in both senses of the word, on almost every page. Our opinions are swayed when we read the results of statistical tests on the safety of nuclear power plants, the popularity of political candidates, or the effectiveness of widely used drugs and vaccines. In these and similar cases, our choices may be influenced; but they are, after all, still our own choices. In many other cases, decisions which gravely affect our lives are made on the basis of statistical studies that we may not even know exist and that most of us would not be able to understand if we did. Government agencies use statistical techniques to justify their budgets and support their claims of effectiveness. The safety of every additive in the foods we eat is determined on the basis of statistical analysis of animal experiments. Where statistical studies are not used to make decisions, they are often used to support or attack decisions, often with each side presenting its own conflicting analysis.

In this climate it is essential for every citizen with even a passing interest in what goes on in the world around him to have some knowledge of statistics. That does not mean having the expertise to compare two highly sophisticated statistical studies and to be able to judge which is more accurate. But it does mean breaking down the mystique and understanding enough about statistics to have a general sense of what this amazing science can and — maybe more important — *cannot* do. It means being able to pick out the more blatant misuses of statistics in advertising and political life. It means having a minimum understanding of what a scientist is doing when he uses statistics to "prove" or disprove a theory. And finally it means understanding the relationship between statistics, a branch of the most exact of all sciences, and the endless complexities and uncertainties of life.

Statistics for Social Change

Chapter 1

STATISTICS AND LOGIC

PART A: Arguments, Deductive and Inductive

The methods and formulas of statistics are not very interesting in themselves. They're not very important either. What makes them both interesting and important is the way they are used to form opinions or make decisions. It is crucial that, when we study statistics, we understand the context in which the methods we learn are to be applied. In everyday life, as well as in scientific research, we find statistics being used in support of statements that are meant to be convincing. Here are two examples.

Example 1: The birth control pill prevents pregnancy in 100% of the cases in which it is used as directed.
During the past year, Mary has been using the pill as directed.
Therefore, Mary is not pregnant.

Example 2: In 1956, 132 women in Puerto Rico were given a particular birth control pill for a period of one year.
None of them dropped dead.
Therefore, it is safe for women to take this pill.

These are examples of arguments that use statistics in support of statements we are meant to accept. But before we can reach any conclusion about them, we must define some terms. First of all, what do we mean by argument? Here we will not be using the word argument in the usual sense of a disagreement or an exchange of opposing views. Rather, we will use it in a very technical sense. By **argument**, we mean a group of statements, one of which makes a point that is being advanced and the rest of which are meant to give support to that point — to convince us of its truth or correctness. A **statement** is any sentence which makes a factual claim.

3

That is, it says something which can be judged either true or false. "The grass is blue" and "snow is white" are both statements, while "get out of here" and "let's go to the movies" are not. Each of the sentences in Examples 1 and 2 are statements.

Statements which make the point of which an argument is trying to convince us are called **conclusions**. Each argument has one conclusion. All the other statements making up the argument are called **premises**. In Example 1 above, "Mary is not pregnant" is the conclusion of the argument. "The birth control pill prevents pregnancy in 100% of the cases in which it is used as directed" and "During this year Mary has been using the pill as directed" are the premises.

Exercise: Which statement is the conclusion in Example 2? What are the premises?

In these examples, it was easy to pick out the premises and conclusions because they appeared in separate sentences with the premises listed first and the conclusion last. This is not always the case in real life. Suppose the first argument had been presented like this:

Example 1a: I'm sure that Mary is not pregnant because people who use the pill properly never get pregnant and Mary's been taking it faithfully for a year.

Although the order has been changed, and the wording is more informal, the argument can still be analyzed as follows:

CONCLUSION: I'm sure that Mary is not pregnant.

PREMISES: People who use the pill properly never get pregnant.
Mary's been taking the pill faithfully for a year.

Exercise: What are the premises and the conclusion in the following version of Example 2?

Example 2a: The birth control pill was tried on a lot of women in Puerto Rico. It must be safe since none of them died.

Being able to sort out the premises from the conclusions of an argument is only the first step in analyzing it. Ultimately we want to be able to say whether the argument has merit or not. And that is where logic comes into the picture. Logic is the study of reasoning. It is the science which clarifies the distinction between good and bad arguments, setting criteria which must be met for an argument to be considered strong, and displaying the flaws which make an argument weak or even worthless. In order to decide whether an argument is good or not, we must consider two quite different things. First we must be concerned about whether or not the premises are true. If the premises are false, the argument cannot hold up. The second point is that even if the premises are true, we must consider whether they actually do lend support to the conclusion. We could construct an argument in which all the premises were true but nevertheless lent no support to the conclusion. For example:

Example 3: Many women use birth control.
Some men are architects.
Therefore, the weather will be good tomorrow.

Both premises are true in the above argument, but they have no connection with each other or, more importantly, with the conclusion, so they cannot lend support or act as evidence for that conclusion.

Logic is not concerned with the first point: the truth of the premises. In order to judge that, we must go to whatever factual sources are available to us. Logic is concerned only with the question of whether or not the premises, if they are true, actually lend support to the argument.

In logic, we use the words "true" and "false" only when we are talking about statements. Every statement is either **true** or **false.** It describes reality accurately, or it does not. In the case of any particular statement, we may not know which is the case. For example, we may not know whether it is true or not that "There is life on Mars." But that does not change the fact that the statement is either true or false. That is, it either corresponds to the actual state of affairs on Mars or it does not.

Arguments cannot be true or false.

Arguments can be characterized as being either **strong** or **weak.** The very strongest kind of argument would be one where the truth of the premises would guarantee the truth of the conclusion. A weak argument would be one which left considerable doubt about the conclusion even if all the premises were known to be true. Before we can say any more about how to tell whether an argument is strong or weak, we must look at the two different kinds of arguments which logicians recognize, deductive and inductive arguments.

Deductive Arguments

Deductive arguments have two important characteristics that distinguish them from inductive arguments.

1. The strength of a deductive argument is an all-or-nothing kind of thing. Either the truth of the premises guarantees the truth of the conclusion, or it does not. If the truth of the premises does guarantee the truth of the conclusion, then the argument is called **deductively valid** (or just valid). This is the strongest type of argument. If it is possible for the premises to be true but the conclusion still to be false, then the argument is **invalid** and therefore worthless.

2. The information provided by the conclusion of a deductive argument is already present in the premises, so no new knowledge is presented by such an argument.

We have already pointed out that it is not the business of logic to determine whether or not the premises are true. For the purpose of deciding whether or not an argument is valid, we assume that the premises are true. We ask ourselves, "If the premises of this argument were true, would the conclusion also have to be true?" If the answer is yes, then the argument is deductively valid; otherwise it is not. Notice, therefore, that **truth** and **validity** are quite different things. It is possible to have a valid argument with false premises. Consider the following examples.

Example 4: Some users of birth control pills have been shown to suffer serious side effects.
All users of birth control pills have been women.
Some women have been shown to suffer serious side effects from the use of birth control pills.

Example 5: If Mary takes the pill, then she will not be pregnant.
Mary takes the pill.
Therefore Mary will not be pregnant.

Both of these arguments are valid. That is, if we accept the premises as true, then the conclusion must also be true. Notice, however, that while the premises of the first argument are true, to the best of our knowledge, the premises of the second are not. It is quite possible for the first premise of Example 5 to be false, especially if Mary takes the pill sporadically. We have no way of knowing whether the second premise of that argument is true or not since we do not even know who Mary is. Nevertheless, the argument is valid, since we see that if the premises were true, then the conclusion would also have to be true. We can think of no condition or set of circumstances that would allow the premises to be true but make the conclusion false. Compare this state of affairs with the following argument:

Example 6: If Mary is taking the pill, she will not become pregnant.
Mary is not pregnant.
Therefore, Mary must be taking the pill.

We can see immediately that even if the premises are true, the conclusion does not necessarily follow. It is certainly possible for Mary not to take the pill and not be pregnant. Perhaps Mary is eight years old, or eighty; perhaps she has been dead for two centuries. Perhaps she doesn't engage in sex, or has sex only with women, or uses a form of birth control other than the pill. In any of these cases, the premises could be true but the conclusion would still be false. Since the truth of the premises does not guarantee the truth of the conclusion, the argument is invalid.

Exercise: Here are two more examples. See if you can tell whether the arguments are valid or not.

Example 7: Mary either takes the pill or is pregnant.
Mary does not take the pill.
Therefore Mary is pregnant.

Example 8: If Mary thinks the pill is unsafe, she will not take it.
If Mary is opposed to contraception, she will not take the pill.
Mary does not take the pill.
Therefore, either Mary thinks the pill is unsafe or she is opposed to contraception.

Formal logic is the study of the rules of inference. That is, it provides the basis for making judgments about the validity of deductive arguments. We will not deal with formal logic here. However, we can still judge the validity of most arguments we will encounter by considering whether or not we can find an example which makes the premises true but the conclusion false. If we can find such an example then we know that the argument is invalid. Not being able to find such an example does not prove that the argument is valid — we may simply have missed it — but it will be the best indication we will have.

Hopefully, then, you recognized that even though the first premise of Example 7 is likely to be false, the argument is quite valid. That is, if it is true that there are only the two mutually exclusive possibilities — the pill or pregnancy — then if you reject one you must accept the other. There is no possible state of affairs in which the premises are true and the conclusion false. In Example 8, the premises are quite reasonable, but the argument is invalid. There are many possible cases that make the premises true and the conclusion false. Note that the premises do not state either that Mary thinks the pill is unsafe or that she is opposed to contraception — only that if either of these were the case, then she would not take the pill. The third premise tells us that she does not take the pill, but the reason for this could be that she wants to become pregnant, is already pregnant, prefers another form of contraception, or any one of a number of other possibilities. So although the conclusion might be true, it is not necessarily so; the premises do not guarantee it. The argument is invalid.

Ultimately what we want to know when we are confronted by an argument is whether or not we should accept the conclusion. If we are talking about deductive arguments, we should accept the conclusion only when we are convinced of both the truth of the premises and the validity of the argument. It is perfectly possible to have a valid argument in which we do not accept the conclusion because we do not believe one or more of the premises. If in fact it turns out that Mary, in Example 1, did not use the pill as directed (making the second premise false), then the conclusion may very well be false. But this in no way affects the validity of the argument. Truth and validity are quite separate conditions, with truth refer-

ring to the content of the premises, and validity to the form of the argument. We must have both truth and validity in order to be guaranteed the
truth of the conclusion.

Let's look at two final deductive arguments to see whether we should
accept their conclusions.

Example 9:   All unemployed women are single.
             Mary is not single.
             Therefore Mary is not unemployed.

Example 10: Some women are employed.
             Some women are married.
             If Mary is a woman, she may be neither employed nor mar-
             ried.

The first argument is valid. If it were true that all unemployed
women are single, then not being single would automatically exclude any
woman from being unemployed. But that premise is not true, so we reject
the conclusion. The premises of Example 9 are both true and the argument is valid (anything can be said about Mary's employment or marital
status since the premises only include some women) so even though the
conclusion gives us very little information, we accept it.

Logicians have worked out the rules of valid deductive arguments in
great detail. We will not be studying these rules here. In examining the
context in which statistics is used, we will find that we will encounter inductive arguments much more frequently than deductive. We will also see
that most of the deductive arguments which are encountered in everyday
life are simple enough so that their validity can be decided by careful reading and thoughtful consideration. We search out fallacies by trying to
imagine cases in which the given premises are true and the conclusion is
false. If we are unable to come up with such a case after careful consideration, we will assume that the argument is valid. Then if we are also convinced of the truth of the premises, we will accept the conclusion.

Before moving on, let us look again at the second characteristic of
deductive arguments: that the information contained in the conclusion of
a deductively valid argument is already present in the premises and that
therefore no new knowledge is gained. If we look back at the examples of
deductive logic above, we see that in each case the conclusion simply states
something already implicit in the premises.

Consider the first example. The conclusion tells us that Mary is not
pregnant, but the first premise says that in 100% of all cases in which the
pill is used as directed there is no pregnancy and Mary is simply one of
those cases, so no new knowledge is gained that was not already contained
in the premises. Or look at Example 4. The conclusion tells us that some
women have had serious side effects from the use of birth control pills, but

those women are just some of the users mentioned in the first premise, so again nothing new is added.

Exercise: Look at Example 9. Explain how the information in the conclusion is already contained in the premises.

Deductive arguments, then, have these two related characteristics: that if the premises are true and the argument valid, then the conclusion is certain to be true, but on the other hand, the conclusion does not add anything new that is not already contained in the premises. If we want an argument that takes us beyond what the premises guarantee, we must look to induction, but then we pay the price of reduced certainty.

Inductive Arguments

We can begin our discussion of inductive arguments by looking at Example 2 again.

Example 2: In 1956, 132 women in Puerto Rico were given a particular birth control pill for a period of one year.
None of them dropped dead.
Therefore it is safe for women to take this pill.

If we compare this argument to the deductive argument of Example 1, we notice two important differences. First of all, the claim made by the conclusion goes far beyond any facts contained in the premises. It generalizes from 132 women to all women; from not dropping dead to being safe. Secondly we note that the truth of the premises in no way guarantees the truth of the conclusion. As we pointed out earlier, only deductive arguments can be so strong that the truth of the premises can guarantee the truth of the conclusion. Inductive arguments vary from very strong, that is, a high probability that if the premises are true the conclusion will also be true, to weak arguments where the truth of the premises only lends a small amount of support to the conclusion. The above argument is not a very strong one. We can see that there are a number of ways in which the premises could be true and the conclusion false. It happens to be true that a particular pill was tested in Puerto Rico as stated. It is also true that none of the subjects of the experiment dropped dead. But should we accept the conclusion? Of course not. The most obvious problem is concluding that the pill is "safe" because none of the subjects died. What of other damage the pill might do, short of killing the subject? But do we even know that the pill will not kill people, perhaps over a longer period of time, perhaps in a different environment? How do we know that if it were tested on more women, it would not prove dangerous to those with health problems different from those in the original test? We could come up with many more objections. The argument is obviously weak.

Let's look at a stronger argument for comparison:

Example 11: A form of birth control pill is tested on 5000 women with no serious health problems who are between the ages of 20 and 40.

During a period of five years, none of them develops a serious health problem.

This pill does not produce serious problems in healthy women between the ages of 20 and 40 over a five year period.

This argument avoids some of the fallacies of Example 2 by more clearly defining what it is talking about and by not generalizing beyond the conditions of the experiment. We can call this a strong inductive argument. Nevertheless, we are not *forced* to accept the conclusion. For one thing, no matter how many examples we find of a particular thing (in this case, 5000 women who suffered no ill effects from the pill) there is no guarantee that if we had examined more cases — 5001 or 50,000 or 5,000,000 — we would not have found one or more who did suffer ill effects. That is the basic uncertainty that we must live with in all inductive arguments.

No matter how strong, inductive arguments can never be 100% certain. This is because they are constructed by looking at a lot of actual cases and then trying to make a true general statement about them. We can look at any number of crows we like — a hundred, a thousand or ten million — and find that they are all black. But that is no guarantee that a white crow will not turn up somewhere sometime. So the statement that we looked at a million crows and they were all black may well be true, but it does not *guarantee* the truth of the conclusion: "All crows are black." The white crow, if it turned up, would be enough to discredit even the strongest inductive argument which concluded that all crows were black. All it takes to discredit an inductive argument is one counter example. **A counter example** is any example which contradicts the conclusion of an inductive argument.

The example of the white crow is a typical one taken from textbooks on formal logic. In real life, we are seldom interested in cases as simple as black and white crows, or things which are always or never true. We can all live very comfortably with the conclusion that crows are generally black, whether or not the mythical white crow remains mythical or actually turns up. On the other hand, consider again Examples 2 and 11. In the case of the safety of birth control pills, we have come to learn of many counter examples: birth control pills have been found to produce serious problems even for some healthy women. Furthermore, any counter examples of this nature have much more serious consequences than does the appearance of a white crow.

Why did it take so long for the dangers of the pill to become known? One reason is a problem inherent to inductive logic, especially when applied to complex scientific phenomena like the effects of chemicals on the human body. Quite a few counter examples may appear without seriously weakening an argument. They may be due to irrelevant factors such as experimental error, unforseen circumstances not built into the experiment, or factors which the observer missed. Therefore researchers may not immediately see the importance of some counter examples which should be taken seriously.

There are other reasons, however, which have more to do with society than with logic. Drug companies were anxious to get their new product on the market. The male-dominated medical profession has always been rather high-handed in its treatment of women's health issues. Population planners wanted a simple, effective contraceptive device, especially for use in third world countries. All of these factors led to tests and experiments which were not sufficient, and to the acceptance of arguments which were not really strong enough to justify unregulated marketing of the new product. (Example 2, for instance, is based on a real study carried out by a drug manufacturer in a low-income housing project in Puerto Rico.)

Induction, then, is a method of reasoning which leads to conclusions of various degrees of probability rather than to certainty. It has the value of taking us beyond what we already know and so moving knowledge forward. But it must be applied carefully, with as much objectivity and concern for the real world effects of our conclusions as possible.

Deduction can never tell us anything new, since the conclusion cannot contain anything that was not there in the premises. Deduction allows us to look at a group of statements (premises) and tease out all the implications contained in them. We tend to start with the premises and work toward a conclusion, which may or may not surprise us — usually not.

Induction works quite differently; we usually start with the conclusion, or rather with an idea we want to prove. We call this initial idea a hypothesis. A **hypothesis** is simply any statement which we would like to prove. In Example 2, the hypothesis was that the pill is safe. Once we have clearly stated our hypothesis, we gather or examine evidence which will support or discredit it. As in deductive reasoning, we must be concerned both with the truth of the evidence (the premises) and with the form of the argument; but no matter how good both are, we will never be absolutely certain of the truth of the hypothesis. Our conclusion will take the form of either accepting or rejecting the hypothesis, but we will only be able to say that we do so because the hypothesis seems more or less probable, not because it is certain to be true or false. Here is an example of an inductive argument:

Example 11:
HYPOTHESIS: The majority of voters prefer Candidate Jones to Candidate Smith.

EVIDENCE: We take three polls to collect voter opinions.
                 The first poll shows Candidate Jones leads Smith 55%
                 to 45%.
                 The second poll shows Jones leading 70% to 30%.
                 The third poll shows Jones leading 62% to 38%.

CONCLUSION: Accept the hypothesis.

This is a perfectly straightforward example where the argument is strong,
assuming that the three polls meet certain criteria. For example, if the
polls represent three people, each going out and asking their ten best
friends, we don't have much in the way of evidence. In order for a poll to
be meaningful, it must cover a large and representative group of people, it
must ask the right questions in the right way, tabulate the results accurately, and so on. But let's assume for the moment that all three polls meet
our standards and that we accept the premises as true and the argument as
strong. Does this guarantee that Jones will win the election? Of course not.
The most we can say is that at the times the polls were taken, Jones'
chances seemed better than Smith's.

Example 12:
HYPOTHESIS: Substance A causes stomach cancer in people.

EVIDENCE: Study 1 shows that 6% of people who eat large quanti-
                 ties of food containing Substance A get stomach cancer.
                 Study 2 shows that in a population where Substance A
                 is unknown, there are fewer cases of stomach cancer.
                 Study 3 shows that 50% of rats fed large doses of Sub-
                 stance A get stomach cancer.

CONCLUSION: Accept the hypothesis.

Example 12 raises a great many more questions than Example 11,
and you may want to think about them before you go on.

Statistical reasoning is simply one form of inductive reasoning. What
we do in inferential statistics is to construct inductive arguments much like
those above. Then we use mathematical techniques to estimate the probability that our experimental results do, in fact, confirm our hypothesis. We
will be dealing with the problems of determining the strength of inductive
arguments throughout the course of this book.

PART B: Falsehoods and Fallacies

In this section, we will look at some of the ways people misuse statistics either intentionally or otherwise, not by calculating the statistics incorrectly, but by wrong uses of statements or arguments. In the language of
this chapter, we will look at various kinds of false premises and invalid
deductive and inductive arguments involving statistics.

False Premises

How can we tell whether a given premise is true or false? Well, let's face it, most of the time we can't be absolutely certain about anything. But we do have certain tests that we use, more or less unconsciously, to help us decide whether or not to believe any given statement. Here are some of the techniques we use in everyday life which will also help us decide about the truth of the premises in logical arguments.

Personal Experience. We can test the truth of some statements by personal experience. If someone says, "It is raining out," we can usually check by looking out the window. This criterion for truth is not always as simple or dependable as it sounds. No doubt you can distinguish rain from someone spraying your window with a garden hose, and you can tell the difference between a black dog and a white dog, but in many cases experience is an inconclusive guide. Most people do not have the experience to decide if the bacteria observed under the technician's microscope actually were streptococcus as reported or not. More and more things we must concern ourselves with are simply beyond our experience.

It is also true that people's subjective experiences differ. You may say, "This dish is too salty," while I might find it just right. Then there is the problem of weighing one person's objective experience against another's. I find that boric acid gets rid of cockroaches, and you find that it doesn't. But despite its limitations, personal experience can be a real help. If you've tried six or seven brands of aspirin and found them all to have pretty much the same effect, you're likely to be somewhat skeptical about an argument that tries to convince you to buy Brand X because it is "50% more effective." So although personal experience isn't likely to provide conclusive evidence, don't hesitate to use it as part of your assessment of a premise.

Source. How many arguments have you heard, punctuated with the phrase, "Sez who?" What may be only an aggressive challenge in a barroom makes good sense in weighing the truth of a statement. We are more likely to believe a figure from the Bureau of Labor Statistics than one from an informal poll. We are more likely to accept as true something read in the *New York Times,* which bases its appeal on its image of accuracy, than something from a sensational tabloid which is primarily concerned with catching the reader's attention. But some stories in the *New York Times* will reflect reality much better than others; we must ask, for instance, whether the "highly placed diplomatic source" who fed an international story to the *Times* reporter is someone whom we trust to tell us the truth. Sources of all sorts, furthermore, often have some self-interest which leads them to provide us with certain statements. A statement about the health of the economy by a President running for re-election, or about off-shore drilling by an oil company, deserves an extra bit of scrutiny — not because the source is ill-informed, but because the source may have a lot to gain by lying.

The source alone can never be the basis for accepting or rejecting the truth of a statement, but examining the source gives us one clue to the statement's value. We should ask ourselves the following questions about the source:

1. What is its previous record for truthfulness and accuracy?
2. How well informed is it likely to be on this particular topic?
3. What self-interest is it trying to advance in making the statement?

Again these are not foolproof tests for detecting error. A careful balance must be found between judging the truth of a statement entirely on the basis of its source and ignoring its source completely.

Common Sense. Finally, one must always rely on one's common sense. That usually boils down to asking how well this statement fits in with other things one knows or accepts as true. Is it reasonable? Does it make sense? No matter how impeccable the source, or how little our personal experience in the matter, we would be unlikely to accept the statistic that Americans consumed 150 billion pounds of sugar last year. Since our population is somewhere in the neighborhood of 200 million, a little arithmetic shows that for the statement to be true, every American man, woman and child would have to consume more than two pounds of sugar every day. Common sense tells us that the figure must be wrong. Of course one seldom encounters such extreme examples, but common sense will often catch accidental (as well as not so accidental) errors on a large scale. Yet again we must take care: Common sense can be misleading. Sometimes we are in danger of rejecting a new truth because it seems to run counter to our common sense. There are some people who still refuse to believe that the earth is round. Common sense tells them that if it were, the people underneath would fall off.

We see that determining the truth of a premise is far from simple. Besides the three general principles just stated, it is also helpful to be aware of some of the kinds of flaws that are likely to appear when statements are put together into arguments that are meant to persuade. Here are some specific and common examples of false premises.

Unclear Definitions or Classifications. No statement can be judged as true or false until we are clear what it actually says — until we know exactly what we are talking about. Take the example of unemployment statistics. What is meant by "the unemployed?" Anyone who is not working, or only people actively looking for jobs? How about those who have given up in despair? Unless we know exactly who is being counted by the government when we read that unemployment has risen to 9% or fallen to 8.7%, we really don't have a fact we can judge as true or not. Here are some examples of ways in which unclear or misrepresented categories have been used to give the public wrong impressions.

Example 13: During the Vietnam War, weekly casualty lists were published every Thursday. Most people assumed these to be more or less accurate accounts of the number of Americans who had died in Vietnam that week. An article in the March 8, 1976 issue of *The New Yorker* tells how one family discovered the truth. Their son had accidentally been killed by American gunfire. Like all the other deaths which occurred by accident, disease, murder, drug abuse, etc., his was not counted in the weekly casualty lists.

Example 14: Presidents Johnson and Nixon both covered huge deficits in their budgets by adding in social security trust funds, which had no business appearing in the budget at all. I.F. Stone, in his May 5, 1969 issue of *The I.F. Stone Weekly*, had this to say about it: "If a present day conglomerate had among its subsidiaries an insurance company, and set up its over-all annual accounts in such a way as to make it appear that the legal reserves of its insurance company were available to meet a deficit in its other business operations, that conglomerate would soon find itself hailed into court by the SEC or its own stockholders. Nixon's claim of a $5.8 billion surplus is just such a fiction. . . . Johnson did the same thing in January when he unveiled his 1970 budget and claimed a surplus of $3.4 billion. The main source of both claims to a surplus lay in the huge social security funds, the government's public insurance business. These and other trust funds are segregated by law and not usable for any other purpose."

Misleading use of definitions and categories is not an easy fallacy to recognize. It helps to be constantly aware of the possibility and to ask questions about whatever categories are used. To find the answers to those questions often means doing more or less difficult research.

Exercise: Here are some figures about the 1977 budget quoted from *The New York Times* of January 22, 1976. If you wanted to know how much the government was really spending on military capacity as opposed to social welfare, what questions would you like answered?

Example 15: OUTLAYS BY FUNCTION

	1977 estimate in millions of dollars
Military	101,129
International affairs	6,824
General science, space, and technology	4,507
Natural resources, environment and energy	13,772
Agriculture	1,729
Commerce and transportation	16,498

Community and regional development	5,532
Education, training, employment, and social services	16,615
Health	34,393
Income Security	137,115
Veterans benefits and services	17,196
Law enforcement and justice	3,426
General government	3,433
Revenue sharing and general purpose fiscal assistance	7,351
Interest	41,297
Allowances	2,260
Undistributed offsetting receipts	−18,840
Total outlays	394,237
Budget surplus or deficit	−42,975

Inappropriate Measurements. Once we are sure that we know the category we are talking about, we are faced with the problem of measurement. Since statistics by definition deals with numbers, those numbers must come from somewhere. We will use the word "measurement" loosely to include all forms of collecting numerical data. Thus "measuring" could include counting people as they entered a building or adding up expenses in a ledger. It could be weighing rats or recording temperatures. It could be examining test scores or assigning ranks to the performers in a music competition. In all these cases, we must consider three problems: what techniques are used to do the measuring, what is being measured by these techniques, and finally, what does the resulting measure actually mean.

Looking first at the techniques of measuring, we can see that some things are easy to measure. Once we have clarified the categories in a proposed budget, we simply add up the number of dollars in each category to measure the total expenditure. Techniques of measurement become a problem when we are measuring something less tangible than money.

Example 16: Suppose you are trying to measure the relative amounts of activity of boys and girls. Eleanor Maccoby, in *The Psychology of Sex Differences* talks about the use of various measuring techniques: "The problems with these methods are seen in the study by Loo and Wenar (1971) — actometers that recorded the amount of gross motor movement a child engaged in did not show boys being more active than girls, whereas teachers reported concerning the same group of children that the boys were more active." (p. 175)

In other words, when two different techniques of measurement were used, opposite conclusions were reached. It seems clear then that if you really want to measure the amount of gross motor activity in children, teacher ratings are not an appropriate technique. On the other hand, teachers' ratings might reflect other aspects of activity, like noise level or fidgeting for example, that actometers miss.

The problem of what is being measured can be seen clearly in the ongoing IQ controversy. Our society suffers from an obsessive need to measure intelligence, yet no two people agree on what intelligence is. But whether one uses the silly definition "intelligence is what IQ tests measure," or a more sophisticated analysis of the factors of intelligence, it is clear that if a test is to have any meaning, it must be measuring the same entity in any group that takes the test. Yet this seems to be the one thing intelligence tests have not been able to do.

Example 17: As N. J. Block and Gerald Dworkin point out in their article "IQ: Heritability and Inequality" in *Philosophy & Public Affairs,* Summer 1974: "Standard IQ tests are without any doubt highly culture loaded. In our view, they are also clearly culture biased in that they require knowledge of, for example, literary, musical, and geographical facts which are differentially available to people with different sociocultural backgrounds." So unless your definition of intelligence includes having miscellaneous literary, musical, and geographic facts at your fingertips, you face a measuring problem.

The question of measuring intelligence by means of IQ, that is, by a number which represents a person's intelligence, raises another problem with measurements. We have come to take it for granted that everything can be measured numerically. When you stop to think of it, it seems fairly outrageous to assume that a single number can tell us very much about something as complex and complicated as a person's mental abilities, if indeed that is what we mean by intelligence. Our willingness to believe that everything can be measured numerically leads to such absurdities as rating the success of a marriage on a scale of one to ten, or claiming a drug is 45% more effective. It may be possible to count the number of quarrels or compliments, the number of kisses or beatings in a marriage, but it would take a great deal more than any combination of all possible numerical tabulations to rate the success of a marriage. Similarly, you may be able to test the speed with which a drug works, you may be able to measure the amount of pain relief various patients report, and you may be able to come up with some measure of side effects, but a numerical value measuring effectiveness (and compared to which other drug or drugs?) doesn't make much sense. A final example from *The World Almanac & Book of Facts,* 1974:

Example 18: "The National Wildlife Federation's 4th EQ [Environmental Quality] Index stands at 54.4. (100 being the ideal environment.) It is down from the 55.5 record by the 3rd Index in 1971." (p. 491)

Exercise: What would we have to know about these numbers to tell us anything about changes in the environment? How much could we learn from one number?

Meaningless Statistics. There are other kinds of meaningless statistics besides those that measure intangibles numerically. One form of meaningless statistic is the kind that someone must have made up to sound impressive because there is either no way the fact could be known or it would take more time and trouble to discover than anyone would be likely to spend.

An example of this kind is amusingly recounted by Susan Trausch in the *Boston Evening Globe* of May 24, 1976:

Example 19: "Somehow I didn't think the National Hot Dog & Sausage Council would hand out baloney. When their news release came in announcing that America was planning to consume 218 million hot dogs this Memorial Day weekend, and that we had put away 16.5 billion, or 77 hot dogs per capita in 1975, I got pretty excited." The rest of the column details her efforts to trace down the origin of these statistics, beginning with calls to the company that put them out and continuing to every research firm they claimed to have used. Needless to say, the source of these statistics was never found.

Rhetoric. The kind of meaningless statistic just quoted can be seen as a form of empty rhetoric, used to make a statement more impressive. Another kind of false statement to look out for is the opposite kind, where the statistics are actually quite accurate (as far as one can judge) but the rhetoric disguises their real meaning.

Example 20: From the *Boston Evening Globe,* August 3, 1973: "The Administration has been pleasantly surprised by the unemployment rate, which most economists expected to increase as production declined throughout the first half of the year. Instead, after rising rapidly from 4.6% last October to 5% in January, the rate has leveled out. . . . The Department said the rate inched up from 5.2% to 5.3%."

Call it "inching up" if you like, claim to be "pleasantly surprised," but the fact is, more people continue to be out of work each month.

Invalid Arguments

In Part A we distinguished between inductive and deductive arguments, but we must also realize that both types of arguments have something very important in common. The premises must support the conclusion, or the argument is worthless. Flaws in logical arguments are called **fallacies** and in this section we will begin by looking at some fallacies that can be found in inductive and deductive arguments alike.

Hidden Assumptions. An **assumption** is any statement that is accepted as true without proof. So assumptions are things people take for granted and don't think very much about. Some of our assumptions are about the way things work in the physical world: "If I let go of an object I am holding, it will fall to the ground." Some are about the way people behave: "If I insult a stranger, she is likely to become angry," or about how people should behave: "Everyone knows that cheating on exams is wrong." Every day we make countless scientific, moral, economic and psychological assumptions. If we didn't, we'd be so busy figuring everything out all over again on every occasion, we'd never be able to function. In a logical argument, we have to make some assumptions because if we had to prove everything, there would be no place to begin. So assumptions are a necessity in logic as in real life. The trouble comes when we are unaware of assumptions which are being made as part of an argument and which simply aren't true, or which may be acceptable to some people and not to others. For example, two people can look at statistical data on the workings of a particular prison program and if one of them assumes that the purpose of prisons is to rehabilitate prisoners and the other that the purpose is to punish, then they are likely to draw very different conclusions from the data. Remember that assumptions, whether stated or not, are always part of the premises of an argument. So if you cannot accept an assumption as true, then you have no basis for accepting the conclusion. Let's look at some examples.

Example 21: An ad for *Famous Grouse* Scotch whiskey, which appeared on p. 75 of *The New York Times Magazine* of June 6, 1976, contains some very impressive statistics showing that 80 proof whiskey will dilute to 45 proof in 9.5 minutes after ice is added, while it takes 90 proof whiskey (like *Famous Grouse*) 18 minutes to do the same. The text runs, in part, like this: ". . . when your drink has been properly cooled — in 30 seconds to a minute — you achieve what one Scotch connoisseur refers to as 'the ideal sip.' From then on, the Scotch drinker's enjoyment typically runs down hill, as the drink loses its freshness. While there is no way to preserve that fresh Scotch flavor indefinitely, we submit that you can sustain the freshness substantially longer with 90 proof *Famous Grouse.*''

The unstated assumption is that it is the proof of the mixture — the actual percentage of alcohol in the scotch and melted ice at any given time — that determines its freshness. The implication is that a freshly mixed drink of 80 proof scotch would taste less fresh than one of 90 proof which had sat until it was diluted to 81 proof. We leave it to scotch drinkers to draw their own conclusions.

The field of prediction is one in which it is most important to look at all assumptions, stated or not. We have only to look at the alternately overflowing and empty classrooms that city schools are faced with in recurring cycles to realize what a problem this is. Most predictions have as their basic assumption: "If current trends continue . . ." Unfortunately they usually don't.

Exercise: What assumptions do you think lie behind the following prediction?

Example 22: The *Boston Evening Globe* of Tuesday, June 1, 1976 ran an article entitled "Life-span Gap Widens for Sexes." It said in part, "The Census Bureau says women outlive men in America probably because they are naturally sturdier, take better care of themselves and — until recently — have shunned the more rugged male occupations. But even the trend toward similar roles and life styles among men and women isn't likely to change things."

Suppressed Evidence. Even more common than missing assumptions are missing facts. Any argument that attempts to prove a point should include all the relevant evidence. Practically this is, of course, impossible. But when an argument is constructed so that only facts supporting the conclusion are included while all others are suppressed, we have an argument that is fallacious by omission.

Many arguments are being made on both sides of the issue of whether or not to build nuclear power plants. On the one hand is the argument that we must find alternatives to oil and on the other that there are serious risks involved in nuclear power. Here is one argument.

Example 23: In an article in the *Boston Evening Globe* of June 15, 1976, James J. Kilpatrick writes: "The big issue was safety. A neutral observer, having no predispositions, may be inclined to wonder how it got to be an issue at all. These nuclear electric power plants cannot 'blow up.' . . . The risk of a catastrophic 'meltdown' is almost immeasurably small. . . . The prospect of some terrorists' overpowering the guards and stealing the fissionable material is demonstrably absurd; . . . Yes, there is a problem in transporting and disposing of atomic waste, but it is no insurmountable problem. In any event, the risk to the public health and safety of nuclear power is minuscule . . ."

One can only hope that the average reader is better informed. Although there is no agreement as to the extent or nature of the risks involved, Kilpatrick is probably the first writer to question that safety is a legitimate issue. Articles on the subject had appeared in the *New York*

Times Magazine, The Nation, U.S. News and World Report, Newsweek, Time and many others in the first half of 1976. *Time* of March 9, 1976 says, "Last March the world's largest nuclear plant, located at Brown's Ferry, Alabama, was well into the chain of events that could lead to a meltdown after human error caused failure of several key safety systems. On a lesser level, a Northeast Utilities plant in Waterford, Connecticut spilled radiation outside the plant when a steam condenser ruptured. Other nuclear power plants have had to suspend operations for anywhere from weeks to several months as a result of equipment failures." (p. 70)

Another logical fallacy which depends on the suppression of evidence is something logicians call a false dilemma. A dilemma is an argument that basically states that of two alternatives, only one can be true. The classical dilemma is that you can't have your cake and eat it too. A **false dilemma** is an argument that tries to convince you that a certain situation presents a dilemma when this is not really the case.

Example 24: In the Summer 1976 issue of *Dollars & Sense* there is an article about environmental standards. Industry tries to convince the public that the enforcement of environmental standards will result in the loss of many jobs at a time when unemployment is one of the country's greatest problems. You can't have a clean environment and jobs too, they are saying. The article points out a number of flaws in this false dilemma, including this: ". . . the Bureau of Labor Statistics has found that each billion dollars of pollution control spending actually *creates* 67,000 jobs in construction, sanitation, and the manufacture of pollution control devices. In 1973 and 1974, more than one million jobs were created in this way." Suppression of this kind of evidence allows the creation of false dilemmas.

Example 25: During a strike of state employees, the *Boston Sunday Globe* of June 20, 1976 ran an article entitled: "How Massachusetts Salaries Compare." It began: "Massachusetts state employees fare well in average salary compared with employees of most other states, according to negotiators for the Office of Employee Relations."

Exercise: What kind of evidence would you look for before deciding whether or not to accept that conclusion?

Irrelevant Evidence. A number of classical fallacies can be grouped under this heading. The general pattern we are concerned with is an argument in which no logical connection has been established between the premises and the conclusion. Here is an example that speaks for itself.

Example 26: On page 89 of the June 21, 1976, *New Yorker* is an ad for Spalding golf balls. A headline reads "The Results of the

$250,000 Longest Ball Challenge," and then follows a list of six makers of golf balls with the words "no show" after each. The text explains: "Last season, Top-Flight put its money where its mouth is: $250,000 to any of these other leading balls that can beat Top-Flight in a distance test. . . . Judging from the turnout, the other leading balls must finally concede what golfers knew all along: Top-Flight is The Longest Ball." Need we point out that there is no logical connection between whether or not various companies wanted to take part in this competition and which ball will indeed travel the farthest?

Another use of irrelevant evidence is that in which the subject is adroitly changed from one argument to another.

Example 27: Two brands of chicken are fighting for control of the market. Brand A charges that Brand B contains chlorine because during processing the fowls are treated and then cooled in chlorinated water. Instead of answering that charge, Brand B countercharges: Brand A is not really fresh chicken because it is cooled in air at temperatures below freezing. Whether or not the latter charge is true, it has no connection with the issue of whether or not Brand B contains an undesirable amount of chlorine.

Statements which often seem relevant to an argument until closer inspection are comparisons which turn out to compare very different things. One instance is the following comparison of two continents' use of world resources.

Example 28: Suppose we are worrying about the increased demand on world resources. We note that according to the Agency for International Development, developed countries are increasing in population at the rate of .9% per year while developing countries are increasing at the rate of 2.5%, or almost three times as fast. We further note that the present populations of Africa and North America are about the same, between 300 and 350 million. We therefore argue that next year the increased resources used by Africa will be three times the increased resources used by North America. Of course this is a fallacy. Since the average North American consumes many times as much in resources of all kinds as the average African, North America will require more extra resources than Africa, even though its population is increasing more slowly.

Exercise: See if you can point out what is wrong with the logic of the following advertisement:

Example 29: On p. 105 of the July 1976 issue of *Ms* is an ad which reads: "Try to find another vodka with a patent on smoothness. Just try. A challenge from Gordon's, the happy vodka. Gordon's is so smooth, so clear, so mixable. It has U.S. patent No. 3,930,042 to prove it."

Inductive Fallacies

Each of the general types of fallacies mentioned in the last section can appear in deductive or inductive reasoning. Here are three more fallacies, all of which we find particularly in inductive reasoning.

Over-Generalization. The very nature of an inductive argument is to look at specific examples and then to draw a conclusion which applies to a more general group. We call the examples we look at a **sample** and the group we want to extend our conclusion to the **population.** Often this population is a set of measurements rather than a group of people; we will define these terms more precisely later. Our concern now is with the question of how general a population we can include in our conclusion from a particular sample. In Example 2 we saw that the main fallacy was one of over-generalization. Here the effect on 132 Puerto Rican women of taking a pill for one year was generalized to all women everywhere for any length of time. This is an obvious example of over-generalization. But on the other hand, we cannot restrict every conclusion to the actual sample used, or we would never be able to draw any general conclusions and would have to repeat every experiment on every possible subject. That makes no sense either.

Even though there are no hard and fast rules to separate over-generalization from proper inference, there are certain guidelines we can look at. First of all, we should ask, "How does the sample we are using differ from the general population in ways which might affect the results of the experiment?" Our sample may have characteristics which are usually lacking in the population, or it may be lacking characteristics found in that population. The most readily available subjects in most psychology laboratories have traditionally been rats and college sophomores, and many of the results of such experiments have been severely criticized on the grounds of over-generalization. The eye reflexes of a college sophomore are probably not very different from those of any other member of the general population, but when the results of rats running mazes are generalized to children learning to read, one may well wonder. Here is an example of over-generalization in a psychological study.

Example 30: In February, 1975, *Psychology Today* ran an article entitled, "Adult Life Stages: Growth Toward Self-Tolerance." The author writes: "My colleagues and I began our search for adult life phases by observing and recording patients in group therapy at the UCLA psychiatric outpatient clinic." Can we assume that psychiatric outpatients in this particular clinic have just those characteristics which would determine the life stages of the general population?

The second question we want to ask about how much generalization makes sense is: "What characteristics which are important to the experiment are present in the general population but are not included in the sample?" For example, if you are testing a drug that is to be safe for everyone to use, then your sample must either include people of all age groups, or your conclusion must be restricted to those age groups tested.

If you were taking a sidewalk poll of voters, where would you station yourself? Near a suburban shopping center or outside a downtown office building? On a college campus or at a baseball game? Near a taxi stand or in a subway station? You would probably come to the conclusion that no one location would guarantee you a completely representative sample. Though some might be more useful than others, a combination of locations would probably give you the best information.

The problem for anyone using inductive reasoning is to make sure that the specific cases used are really representative of all sections of the general population. This problem will be taken up in greater detail in the chapter on sampling.

Insufficient Evidence. Since the inductive process is one where a number of specific cases are examined in order to reach a general conclusion, there is always a question of how many cases are sufficient. This question will have different answers in different circumstances, as we shall see later. But there are certain kinds of insufficient evidence that should be obvious to us even without technical statistical knowledge.

One type of insufficient evidence that is frequently encountered in the popular press is anecdotal evidence. Here the writer goes into great detail in describing one or more individual cases, thus obscuring the fact that the conclusions are based on very little hard fact, or on a very small number of cases.

Insufficient evidence does not mean evidence that is wrong, though it might also be that, but evidence that by itself cannot justify accepting the conclusion. Individual cases, or **anecdotes** usually don't provide sufficient evidence. Although an anecdote may hint at something worth investigating, it doesn't in itself constitute evidence. Nor, for that matter, do a whole series of anecdotes.

Example 31: An article entitled "Why You're So Tired" appeared in the November, 1975 issue of *Ladies Home Journal.* The article begins with an anecdote about Mary W. who felt tired all the time. After her own doctor decided there was no physical cause and referred her to a psychiatrist, she went instead to see Dr. Goldberg, the author of the article. Dr. Goldberg tells us he diagnosed a thyroid deficiency, and then goes on to say, "As a specialist in endocrinology (the study of hormone and glandular function), I have a number of tired women referred to me. And I take strong issue with the prevailing medical misconception that most are neurotic, bored, sex-starved or grist for the psychiatrist's mill. In my experience, *at least half such women can be shown to have a definite physical problem that causes their chronic tiredness.*" (Italics his) Now much as one may agree with Dr. Goldberg's distaste for prevailing medical opinion, the anecdotal evidence of an endocrinologist is hardly sufficient evidence for the statement that half of all tired women suffer from "a definite physical problem."

Similarly, arguments by analogy are of little value in themselves. An **argument by analogy** simply says that since A is like B and the conclusion drawn from A was C, the conclusion to be drawn from B must also be C. These arguments are popular with politicians who wish to predict the future on the basis of their reading of history. "The fall of Rome was preceded by a relaxation of moral strictures, so all this permissiveness today will lead to certain collapse tomorrow." The difficulty with analogies is that they tend to concentrate on one factor in a very complicated situation, ignoring all the other factors which may have played even more important roles. Recent economic developments are a case in point. Past experience had shown that there was a trade-off between inflation and unemployment, so that one could reduce inflation by allowing greater unemployment or reduce unemployment by tolerating more inflation. The mid 70's have shown that this is not always true. It is perfectly possible to have both high unemployment and high inflation.

Advertisers are particularly fond of using argument by analogy, either by encouraging you to identify with the glamorous endorser of their product (if she plays so well with a Brand X racket, so will I) or by trading on their past reputations.

Example 32: Steinway & Sons runs an ad which shows a letter in old fashioned writing saying, "I have decided to keep your grand piano. For some reason unknown to me it gives better results than any so far tried. Please send bill with lowest price. Yours, Thomas A. Edison." Underneath, it says, "It's still a bright idea to choose a Steinway" (*Harper's Magazine*, December 1975, p. 19).

Exercise: Give two reasons why the evidence isn't sufficient.

All this is not to say that analogies between past and present, between here and there, between animal and human behavior, etc., are always false. Sometimes, indeed, they are very useful in providing clues to a possible explanation or prediction. It is just that by themselves they do not constitute sufficient evidence.

Causal Fallacies. One of the most difficult, and often most interesting areas of scientific research or political speculation is the area of causal connections. Why did this happen? What makes people behave like that? How does this mechanism function? We will look at causal problems in detail in the chapters on correlation and linear regression. Here we will look only at the general problem.

What do we mean when we say that one thing *causes* another? If Event A causes Event B, does that mean that every time we observe Event A, we observe Event B? Suppose Event A is noting a case of typhoid and Event B is finding that the water supply is polluted. Common sense tells us it's more likely that B (the polluted water) caused A (the typhoid) than the other way around. We can test this out by noticing that if we measured the pollution level of the water for a period of time, we would find that typhoid never appeared until the water reached a certain level of pollution. So Event B must happen before Event A in order for us to suspect B of causing A.

Let's see now if that helps us find a definition for causation. Can we say that Event A causes Event B if we always observe Event A before Event B occurs? Not necessarily. For example, if we let Event A stand for "day" and Event B for "night" we have met this condition, but do we have cause and effect? Can we say day causes night because it always precedes it? Of course not; both are caused by the rotation of the earth on its axis. So we need to be sure that if one event always and invariably follows another it is not because both are caused by some third event. This is not always easy to prove, one way or the other, though there are ways of designing experiments, to be discussed later, which help.

So far we have been oversimplifying real causal problems by looking at cases where one event always causes another. In real life, things aren't that simple. We are more often concerned with contributing causes than with sole causes. Most of us are fairly convinced that excessive smoking causes lung cancer. But it is also clear that it is not the only cause of lung cancer, nor does it always cause it. Some people get lung cancer without ever lighting a cigarette, while others smoke several packs a day and die of old age at 90. Cigarette smoking, then, is a contributing cause of lung cancer, not a sole cause.

Finally, we must be aware that it is sometimes very difficult to sort out what is cause and what is effect, since there tend to be series of events in which the effect of one cause can become the cause of the next. It is especially confusing to researchers when these cause-effect relationships form a spiral, or feedback loop: the old chicken and egg controversy. Here is an interesting example:

Example 33: *Ramparts Magazine,* in its August/September 1975 issue, printed an article, "How Poverty Breeds Overpopulation (and not the other way around)" by Barry Commoner. Where most people tend to assume that overpopulation causes poverty, Commoner offers some evidence that the contrary is also true. He quotes a study by a Harvard team of researchers in India who "tested two of the possible approaches: family planning and efforts (also on a family basis) to elevate the living standard. The results of this test show that while the family planning effort itself failed to reduce the birthrate, improved living standards succeeded." (p.24) Overpopulation can be a contributing cause of poverty, but it is important to realize that there is good evidence that poverty can also be a contributing cause of overpopulation.

So the next time you're confronted with an argument which says that A causes B, stop to think whether it's not possible that B causes A instead or that C causes both. And then consider the possibility that the connection between the two events is purely accidental.

Multiple Fallacies

In the preceding pages we have discussed four types of false premises and six types of logical fallacies. These categories should be helpful to you in spotting shaky arguments. But don't expect to be able to place every questionable argument you encounter into one and only one category. The idea of categorizing fallacies comes from formal logic where everything is spelled out and it is easy to label each error exactly. You will find in real life situations that categorizing is not nearly so simple. First of all, you will find that fallacious arguments often contain more than one kind of error. Then, when you look at one particular error, you will find that how you classify the error will depend on how you interpret the logic of the argument.

Look again at Example 2. We have already pointed out that there are several different kinds of errors involved. Let's look at two:

1. There is certainly something wrong with equating "safe" with non-lethal.
2. There is a problem with using 132 Puerto Rican women to represent all women.

Now how do we classify these errors?

1. This could either be an unclear definition (what does "safe" mean?) or a misleading classification (one can't classify a pill as

safe just because it didn't kill anyone) or a hidden assumption (the assumption being that something is safe if it doesn't kill).

2. We can say that the problem here is over-generalization by generalizing from 132 Puerto Rican women to all women everywhere, or we can say that there was insufficient evidence since the sample was too small and not representative of every kind of person likely to take the pill.

The point of all this is that we are more interested in spotting fallacies than in classifying them. An argument is fallacious not because it is an "argument by analogy" but because it presents insufficient evidence and we can point to the insufficiency. We say the implied argument in Example 32 is fallacious not because we recognize it as an example of "argument by analogy" but because we see its weakness: We can ask whether Edison was an expert on pianos, and if so, whether his taste 100 years ago would match our taste today. We can question whether a company that made good pianos 100 years ago (if, in fact, they did) necessarily produces the same quality product today.

As you read books, newspapers and magazines, keep the categories in mind to help you spot fallacious arguments, but be prepared to point out exactly what is wrong with the argument by thinking up counter-examples, seeing specifically how and where the argument is wrong.

SUMMARY

PART A

A **statement** is any sentence which makes a factual claim and can therefore be either true or false. An **argument** is a series of statements consisting of one or more premises and a conclusion. The **conclusion** is the statement which the argument advances, while the **premises** make up the evidence for that statement. Premises may be either true or false. A **true** premise states something which corresponds with our understanding of the state of the world, while a **false** premise does not. An argument may be either strong or weak. A **strong** argument is one in which the premises provide a high probability of the conclusion being true while a **weak** argument is one where the premises provide only a low probability.

Deductive arguments are arguments which are either valid or invalid. A **valid** deductive argument is the strongest kind of argument: if the premises are true, then the conclusion must be also true. An **invalid** deductive argument lends no support to the conclusion. Deductive arguments, even when valid, do not contain new information in the conclusion that is not contained in the premises.

Inductive arguments have conclusions which make statements that go beyond the facts contained in the premises. Inductive arguments range from strong to weak, but even the strongest inductive argument provides no guarantee of the conclusion being true when the premises are true. A **counter example** is an example which meets the conditions of the premises but contradicts the conclusion, thus tending to discredit the argument.

PART B

Some guidelines for judging the truth of a premise are the following:

1. Does it match one's direct experience?
2. Is the source known to be reliable and competent? What is its self-interest?
3. Does it make sense and tally with other known facts or accepted theories?

False premises often have one or more of the following characteristics:

1. Unclear definitions or misleading classifications.
2. Inappropriate techniques of measurement or meaningless numerical values.
3. Statistics which are meaningless because they are unknowable or exact figures that could not really have been collected.
4. Accurate statistics which are hidden by conflicting rhetorical statements.

Invalid arguments can be the result of:

1. Hidden assumptions
2. Suppressed evidence
3. Irrelevant evidence

Invalid inductive arguments can have any of the characteristics mentioned above plus:

1. Over-generalization
2. Insufficient evidence
3. Causal fallacies

Multiple fallacies often characterize fallacious arguments. Any particular fallacy may also be classified in different ways. It is more important to recognize and explain fallacies than to classify them.

EXERCISES

*Answers in the back of the book.

PART A

Read through each of the following arguments carefully.

1. For each argument, identify the premises and the conclusion.
2. Classify each argument as deductively valid, inductively strong, or worthless.

   a. It was impossible, during 1976, for a family of four to be adequately supported by a man working full time at the minimum wage. According to statistics from the Labor Department, it cost $10,041 for a family of four to maintain an austere standard of living. The minimum wage in 1976 was $2.30 per hour.

b. *The swine flu vaccine was tested on 10,000 people. None of them got swine flu. I should get vaccinated since I don't want to catch the flu.

c. On Feb. 1, 1980, the *Boston Globe* reported that a woman earns 59¢ for every $1 earned by a man. According to the *Statistical Abstract* the comparable figure for 1970 was 62¢, for 1973 it was slightly less than 62¢, for 1975 it was also about 62¢. Women's salaries will not equal men's by 1981.

d. *On June 3, there were 24,789 people on the beach in Santa Monica. Since one out of every three people had a radio on, there were 8,263 radios playing.

e. On June 1, 1977, a market research firm shopped in 100 supermarkets in the Greater Boston area. It found that food was cheaper than last year. Prices were 15% lower than on June 1, 1976.

f. *If all hockey players who got into fights were fined, there would be a significant decrease in fights on the ice. Fighting hockey players will not be fined, so no decrease can be expected.

g. No woman is a good mechanic. Any good mechanic can fix this car. No woman can fix this car.

h. *On April 29, 1976, the *New York Times* reported that in the Pennsylvania primary Carter won more votes than any of his opponents among people from all occupations, all age groups except those over 65, all religions except Jewish, and both races. Therefore it was to be expected that Carter would win the election.

PART B

1. *For each argument in Part A, state whether you believe the premises to be true, reject them as false, or have no opinion. Give reasons.

2. For each argument state whether you accept or reject the conclusion. Explain why, referring to exercise 2 Part A and exercise 1 Part B.

3. Read the following three excerpts from the June 14–20, 1975 issue of *TV Guide*. Try to identify the premises and conclusions of each excerpt and then decide about the truth of the premises and the strength of the arguments. Explain to what extent you accept or reject each conclusion and why you do so.

a. From "Violence — as Human as Thumbs" by Isaac Asimov:
    Mankind lived by violence for uncounted thousands of years before history began. There were long ages in which human beings had to kill animals for food. . . . When a city was taken, its inhabitants were quite likely to be killed or enslaved. . . . Under such circumstances it was important to fight to the death. . . . Young-

sters had to get used to violence, had to have their hearts and minds hardened to it. . . . And now it's over! That old-time violence that's got us in its spell must stop! . . . The new enemies we have today — overpopulation, famine, pollution, scarcity — cannot be fought by violence. . . . It is with tales of brotherhood and cooperation that we had better propagandize our children. If we choose not to, if we continue to amuse ourselves with violence just because that worked for thousands of years, then the enemies that can't be conquered by violence will conquer *us,* and it will all be over.

b. From "Does TV Violence Affect Our Society? Yes," by Neil Hickey:

It is virtually impossible for Americans, of any age, to avoid the depiction of violence on their TV screens. (One scientist estimates that by the age of 15 the average child will have witnessed 13,400 televised killings.) . . . Meanwhile, violent crime has been increasing at six to ten times the rate of population growth in the United States. (Obviously, nobody blames all of that on television.)

Dr. Robert M. Liebert, a psychologist at the State University of New York (and a principal investigator for the Surgeon General's report) says unequivocally: "The more violence and aggression a youngster sees on television, regardless of his age, sex or social background, the more aggressive he is likely to be in his own attitudes and behavior. The effects are not limited to youngsters who are in some way abnormal, but rather are found for large numbers of perfectly normal American children." That conclusion arises from analysis of more than 50 studies covering the behavior of 10,000 children between the ages of 3 and 19.

c. From "Does TV Violence Affect Our Society? No," by Edith Efron.

Network TV's economic existence, and an enormous number of its technical calculations, are totally based on the certain, *proved* knowledge that the overwhelming majority of the U.S. public and its kids is fixated on the simple, continuous vision of good, just men. To a striking degree, network TV's profits flourish in the loam of hero worship.

. . . If, indeed the "cumulative" watching of evil is turning us all, gradually, into depraved beings, then the "cumulative" watching of good must be turning us all, gradually, into saints! You cannot have one without the other. That is, unless you are prepared to demonstrate that evil is something like Cholesterol — something that slowly accumulates and clogs the system while good is something like spinach, easily digested and quickly excreted.

Chapter 2

PERCENTS
AND PERCENTILES

PART A: Numerical Descriptions — Comparisons

When we want to describe something, we often find it necessary to use numbers. For example, it is much more accurate to say "the temperature is 85° and the humidity is 90%" than it is to say "it's hot." We have a clearer idea of someone's income if we are told she earns $100,000 a year than if we are told that she is rich. "There's a lot of crime these days" is less meaningful than "there were 48 murders in Greater Boston last year." But often numbers by themselves don't provide enough information to be meaningful. Take the last number for instance. Is 48 murders a lot for a city that size that year, or not? How many murders were there in New York? In the U.S.? If we can make a comparison, the numbers will give us more information.

Comparison by Subtraction

Suppose a loaf of bread sold for 52¢ last year and sells for 78¢ now. We can compare these prices in two ways, by subtracting or by dividing. If we want to focus on the absolute difference between two quantities, we compare by subtraction. So if we want to know how much more the bread sold for this year than last, we subtract:

$$78¢ - 52¢ = 26¢$$

We can then say that the bread costs 26¢ more this year or that it costs 26¢ less last year. The difference between the two prices is 26¢.

Comparison by Division

Ratios. The other method of comparing numbers — comparison by division — is often more useful. This method tells us how many times one

33

number is the size of another, or what fraction one number is of another. Continuing our bread example, we can compare the two prices by dividing either into the other:

$$78¢ \div 52¢ = 1\frac{1}{2} \text{ or } 1.5$$

This tells us that the price of bread this year is one and one half times what it was last year. Alternatively, we can divide 52¢ by 78¢:

$$52¢ \div 78¢ = \frac{2}{3} \text{ or } .67$$

That is, the price of bread last year was two thirds of what it is this year.

A comparison by division is called a **ratio.** A ratio can be thought of as a division problem or as a fraction. The ratio of 78 to 52, for example, can be written in the following ways:

$$78:52 \qquad 78 \div 52 \qquad 78/52 \qquad 3/2 \qquad 1.5$$

Exercise: What are five ways of writing the ratio of 52 to 78?

So far we have discussed only one kind of comparison, the kind where we are dealing with two similar quantities: two prices for the same object, the heights of two buildings, etc. But we can also use ratios for another kind of comparison, the kind where we are considering a part of something in relation to the whole. Suppose someone tells us that the federal government spent $84 billion on the military last year. When numbers get that large, they no longer have much intuitive meaning for us. But if we knew the size of the total federal budget, we could see what part of it was being spent on defense. If the total federal budget was approximately $300 billion, then we can form the ratio $84/300 = 7/25$, and see that more than one fourth of the federal budget was spent on the military.

Once we know what part of the federal budget is spent on the military, we can make further comparisons. We might want to know what part is spent on other things, like education for instance. Suppose we find that the expenditure for education was $18 billion. The ratio there would be $18/300$ or $3/50$. If we now want to compare defense spending with education spending, we have two fractions to compare: $7/25$ and $3/50$. Here we see one of the shortcomings of ratios. Fractions with different denominators are hard to compare. We can change to a common denominator ($14/50$ and $3/50$ are easier to compare than the original fractions) but that doesn't entirely solve the problem. Suppose we now want to compare these two ratios to the ratio for transportation. We would have to find another common denominator, this time for three fractions, and so on. A much better way to compare these numbers would be to get rid of the denominators entirely.

Percents. Changing fractions into percents is the way to get rid of denominators. Percents are fractions with a common denominator of 100. Since the denominator is always the same, we need not write it and can use the symbol "%" instead. That is, if we have the fraction $7/25$ we can change it to 28/100. Then we can write 28/100 as 28%. Similarly 3/50 = 6/100 = 6%.

Exercise: In the example given above if $24 billion was spent on interest payments, what percent of the federal budget does this represent?

We see that percents provide a convenient form for making comparisons. With percents we can see at a glance what fraction each part is of the whole and we can also compare parts to each other with ease.

If you have forgotten how to work with percents, you should review these basic skills now:
  1. Changing back and forth between fractions, decimals and percents
  2. Finding what percent one number is of another (30 = ?% of 60)
  3. Finding a given percent of a number   (50% of 60 = ?)
  4. Finding the base, given percent & percentage   (30 = 50% of ?)

When working with percents, it is very important to be aware of the base at all times. The **base** of a percent is the whole quantity of which the percent is a part. In the above example, the base, for both the 28% spent on defense and the 6% spent on education, is the total amount of the federal budget, $300 billion. When we talk about percent, whether we actually say so or not, we are always talking about a percent *of* something. That something is the base. If we don't know, or can't find out what that base is, then the percent itself is meaningless. That is why advertisements such as "20% more effective" or "50% faster" provide no basis for comparing products or services. If we are not told what base is being used we cannot judge the truth of the statement. 20% of what? 50% of what? We must always be in a position to know the answer to such questions (to know what the base is) in order for the percent to have meaning.

Comparing Over Time: Rates of Increase or Decrease

Another common comparison is a comparison over a period of time. Let's go back to our bread example. In January of 1975 the price of a loaf of bread was 52¢. In January of 1976 it was 78¢. What was the rate of increase? The **rate of increase** (or **decrease**) is defined as the ratio between the amount of increase (or decrease) and the earlier quantity. To find the rate of increase in the price of bread, then, we first find the difference between the original price and the new price (subtraction) and then find the ratio between that difference and the original price (division):

$$78¢ - 52¢ = 26¢; \qquad 26¢/52¢ = .5 \text{ or } 50\%$$

Note that rates are usually stated as percents. We can now say that the rate of increase in the price of bread has been 50% in the past year.

Exercise: What is the rate of increase for a change from 20¢ to 25¢ per year?

The time factor of one year is very important here. To see how it operates in calculating the rate of increase, let's look at another example.

Example 1: Suppose eggs were to increase in price by 5% per month for a period of a year. This is what would happen:

Month:	1	2	3	4	5	6	7	8	9	10	11	12	1
Price:	60	63	66	69	72	76	80	84	88	92	97	102	107
Absolute Increase:		3	3	3	3	4	4	4	4	4	5	5	5

At the end of the year, then, the eggs which started out costing $.60 end up costing $1.07. At 12 increases of 5% each, we might expect an increase of 60%, but this turns out to be incorrect. 60% of $.60 is $.36, while the actual increase is $.47 ($1.07 − $.60 = $.47). Where did the extra increase come from? The answer is that although the rate of increase remains constant from month to month, the base on which we figure the increase keeps rising. The first month the increase is figured on $.60, the next month on $.63, then on $.66 and so on. That's why the time factor is so important. Rates of increase or decrease are like compound interest on a bank loan: they have to be re-figured for each unit of time, using the new base at each stage.

In actual practice, of course, it would be very unusual for the rate of increase to remain constant for many months, especially in food prices which normally tend to fluctuate. Newspapers, in reporting inflation trends, try to take this into account in their predictions, and usually report yearly as well as monthly rates. In reading such stories, it is necessary to pay close attention to the wording to see just which time factor is actually being used. There's a huge difference between an inflation rate of 1% a month and 1% a year.

Exercise: How much more would you pay for a 50¢ carton of milk at the end of a year if the rate were 1% per month? 1% per year?

Comparing Rates of Change

Rates can be compared just like any other numbers. If the rate of inflation, for instance, was 12% last year and this year it's 10%, we can compare by subtraction and say that there was 2% less inflation this year than last. That is, there was a decrease in the rate at which prices rose. Notice that a decrease in the inflation rate does not mean that prices are going down; it just means that the rate at which they are rising has slowed down somewhat. If prices actually fell by 10% during a year, we would express that year's rate of inflation as "−10%" (minus ten percent).

Comparing Scores

If you've ever looked at the results of a Scholastic Aptitude Test (SAT) or other standardized educational test, you've seen such terms as "raw score," "percent score," "rank," and "percentile rank." These terms are also used to express the results of tests for intelligence, motivation, and other psychological tests. In order to interpret the results of all these tests, the scores have to be expressed in some way which allows us to make comparisons. This confronts us with some special problems and some special techniques.

Raw Scores. The simplest way to express a score on a test is to count the number of items which were completed correctly (or incorrectly, if that turns out to be more convenient). We call this the **raw score.** Note that the raw score is not very helpful if that is the only information we have. If you say that you got 12 problems right on a math test, we know very little unless we know how many problems there were on the test. A raw score of 12 looks quite different if it is the score on a test containing 12 problems than it does on a test of 20 or 50 or 100 problems. Therefore raw scores are almost always converted to percent scores.

Percent Scores. The **percent score** is calculated by dividing the number of correct items by the total number of items on the test and expressing the answer as a percent. If there were twenty questions on the test and your raw score were 12, the percent score would be:

$$12 \div 20 = .6 \text{ or } 60\%$$

That is, 60% of the problems on the test were done correctly.

While the percent score provides more information than the raw score, it still leaves out some important information. The percent score tells us very little about how one score compares with the rest of the class. Perhaps the test was so difficult that 60% was the highest score. Or 60% could have been the lowest — or anywhere in between.

Rank. One way to compare different people in a group is to rank them, giving the person with the highest score a rank of one, the next highest two, and so on.

Example 3: Suppose 10 people take a test and their scores are:
90%, 70%, 50%, 50%, 80%, 60%, 50%, 70%, 60%, 70%
To rank them, we must first re-arrange the scores in order of size, beginning with the highest score:
90%, 80%, 70%, 70%, 70%, 60%, 60%, 50%, 50%, 50%
It is clear enough that the person who got 90% ranks first and the person with 80% ranks second. But what of the three people who scored 70%? The 70% scores occupy the third, fourth and fifth rank, but since they are all the same, they should have the same rank. So we add up the three ranks and divide by 3, taking the average:

$$(3 + 4 + 5) \div 3 = 4$$

We say that each of the three scores ranks in fourth place
The two 60%'s occupy ranks 6 and 7, so their rank is

$$(6 + 7) \div 2 = 6.5$$

Exercise: What is the rank of the three 50%'s?

This procedure is simple enough, but again we fall into the same trap as with the raw scores. The rank gives us very little information unless we know how many persons were in the group. A rank of 5 might be pretty impressive out of a group of 100, but is a good deal less so out of a group of 10. This difficulty too can be overcome.

Percentile Rank. We approach percentile rank much as we did percent scores except that now we are interested in a percentage of people rather than test questions. That is, we want a ratio between the rank of one person's score and the total number of scores. Again we must be concerned with the problem of several people having the same score. The formula that we use for percentile rank is this:

$$PR = \frac{L + \frac{1}{2}S}{T}$$ where PR = percentile rank $\quad L$ = number of
$\qquad\qquad\qquad\qquad\qquad T$ = total number of scores $\qquad$ lower scores
$\qquad\qquad\qquad\qquad\qquad S$ = number of scores
$\qquad\qquad\qquad\qquad\qquad\qquad$ the same value

In other words, we add the number of people who got lower scores to half the number of people who got the same score and divide by the total number of scores.

Notice that percentile ranks and ranks run in opposite directions. While a low number indicates good performance when we are using ranks, it indicates poor performance when we are using percentile ranks. There is no reason for this; it is simply the way it is conventionally done.

Example 4: Using the same data as in Example 3, what is the percentile rank of someone scoring 70%?
Using the formula above, we see that $L = 5$ because there are 5 scores lower than 70% (2 60%'s and 3 50%'s). $S = 3$ because there are 3 scores of 70%, and $T = 10$ because there are 10 scores in all. So we have:

$$PR = \frac{5 + \frac{1}{2}(3)}{10} = \frac{5 + 1\frac{1}{2}}{10} = \frac{6\frac{1}{2}}{10} = \frac{6.5}{10} = .65 \text{ or } 65\%$$

We say that a score of 70% on that particular test in that group has a percentile rank of 65. That means that 65% of the people who took the test scored lower than 70%, while 35% scored higher.

Exercise: Calculate the percentile rank for 60% in Example 3.

PART B: Percentage Pitfalls

As Part A suggests, percents offer as many opportunities for obscuring the facts as for clarifying them. Some misuses of percents result from mistakes in the actual calculations. The percent simply does not express what it claims to. More frequently, there is nothing wrong with the percent itself, but the conclusions drawn are unwarranted for one reason or another. Let's look at some examples.

Problem Percents

Most of the actual calculation errors in the area of percents are the result of some problems with the base. Often we may think we know what the base is, only to find out that our assumptions are wrong, that the base is not accurately specified, or that we're using the wrong base for the purpose or the time period.

Poorly Defined Base. Although regulations have become tighter in recent years and advertisements which make vague claims and state no base for a percent are becoming rare, the base is still often so poorly defined as to be no help.

Example 5:   In the *NY Times* of Sept. 9, 1976, a magazine called *Modern Bride* tries to convince manufacturers to advertise in its pages because brides buy more new household articles than other potential customers. The ad concludes, "You can't find a better customer anywhere. And you reach her in *Modern Bride.* The one magazine read by eight out of every ten girls passing through the bridal market."

At first reading this sounds as though 80% of all women who get married read *Modern Bride.* This seems so unlikely that if the statement is not an outright lie, then "girls passing through the bridal market" must mean something very different from "every woman who gets married." Since we cannot guess who is included in this vague category, we don't know the size of the base, and therefore the percent is useless. Eighty percent of an unknown number is still an unknown number.

Incorrect Base. Rates are sometimes figured on a base which is simply not the correct one. An example of this is the way interest rates on loans used to be figured.

Example 6:   Suppose you borrow $120 for 1 year. You are required to pay back $11 a month, so you know that your interest charges amount to $12 for the year. ($11 times 12 months = $132; $132 − $120 = $12.) What rate of interest are you paying? If $120 is used as the base, that amount comes out to an interest rate of 12/120 or 10%. But this is incorrect.

The reason the $120 is the wrong base is that you didn't have the use of the $120 for the whole year. During the first month, you had the whole $120. During the last month, you had only $10. On the average you had the use of about $65 ($\frac{120 + 10}{2} = 65$) during the year, so $65 is a more accurate base than $120. That makes the true interest rate 12/65 or 18.5%. The new truth-in-lending laws now force loan companies to reveal this true interest rate, but customers who don't read the fine print and do their own figuring still frequently make this mistake.

Here is another example of the wrong base being used. See if you can tell what is wrong with the following example.

Example 7: During the spring of 1975, a large university received the following letter from its food service company: "This summer the cafeteria price will be $4.50 per person per day. This represents a 17% increase over the rate charged during the summer of 1974. The increase is based on an estimated 6% cost of living labor increase and an 11% food cost increase."

Apparently the 17% increase was arrived at by adding the estimated increase in labor costs to that of the increase in food costs. This calculation is wrong on two counts. The correct way to adjust prices when some factors (in this case labor and food) increase in cost is first to determine what part of the cost of the finished product (in this case cafeteria meals) each factor represents. Since other parts of the meal costs, such as equipment, transportation, etc. did not increase, only that part of the cost which did increase can contribute to higher prices, and it must do so in the correct proportion. So for example if labor costs represented 25% of the cost of a meal and food costs represented 40%, the total increase should be:

$$(6\% \times 25\%) + (11\% \times 40\%) = 1.5\% + 4.4\% = 5.9\%$$

Under no circumstances could the increase be as much as 11%, since the increase in food costs represent only part of the cost of the meal that is served. Adding the 6% to the 11% compounds the error, so that the total increase of 17% is completely indefensible.

False Comparisons. Comparing percents is a very tricky business. Two percents are comparable only if they have comparable bases. If the bases are measuring different things, or are of very different sizes, the comparisons become meaningless. Here is a hypothetical example which illustrates how misleading percents can be when their bases are very different in size.

Example 8: Jones and Smith are candidates in an election. During the period of one month, while they are campaigning, a poll shows that support for Jones has increased by 10% while support for Smith has decreased by 10%. Do we assume that Jones will win? Not if the figures look like this:

Number of votes for:	July	August
Jones	30	33
Smith	3000	2700

There are many possible examples of percents which cannot be compared because their bases are measuring different things. An interesting one arises in the calculation of profits. Corporations like to look as though they were making healthy profits when they are talking to stockholders, while they'd rather have the consumer think they're going broke. The general public can get very different pictures of a corporation's profits depending on how the story is told. Much of the confusion rests on complicated accounting procedures, but there is one very simple issue that is relevant here. Once the actual amount of profit has been calculated, there are several possibilities for expressing the profit as a percent, depending on what is used as the base.

Example 9: In an article entitled "Supermarkets in a Crunch," in *The New York Times Magazine* of Feb. 8, 1976, Allen J. Mayer, though discussing mainly the problems the A & P grocery stores are having, explains about profits in general. Supermarkets complain that they are making a mere 1% profit on sales. In other words, their total profits equal 1% of their total sales. But the percentage profit figures we are more used to hearing — usually in the range of 5% to 15% — refer to profits as a percent of company's assets, not its sales. "In 1975," Mayer reports, "the supermarket industry's average return on equity (one form of measuring assets) was running close to 8% — slightly below average for all industries, but nowhere as bad as the 1% profit-margin figure that grocers use to justify their claims of poverty."

It is interesting to note that *How to Lie With Statistics* by Darrell Huff, which was published in 1954, uses the same example of A & P's profits (they were then 1.1%). Huff concludes his explanation of the difference between percent of sales and percent of investment with the following statement:

"There are often many ways of expressing any figure. You can, for instance, express exactly the same fact by calling it a one percent return on sales, a fifteen percent return on investment, a ten-million dollar profit, an increase in profits of forty percent (compared with 1935–39) or a decrease of sixty percent from last year. The method is to choose the one that sounds best for the purpose at hand and trust that few who read it will recognize how imperfectly it reflects the situation."

This leads us to our final category of questionable uses of percents. Although in some cases it would be questionable to call the selection of the most flattering measure a fallacy, in others it obviously is. Selecting the "best" measure to fit the data is not an easy task. The measure selected should be appropriate to answer the question being investigated, and the assumptions associated with that measure should be spelled out. As the examples given below indicate, picking a flattering measure to support a particular point of view and calling it the "best" measure for the purpose can lead to wrong conclusions.

Faulty Arguments

False Assumptions. To a large extent the statistics that get quoted in newspapers or magazines are the result of certain assumptions which are made, but not stated. These are the most difficult errors to detect, since there is nothing in the statement to point us in the right direction. Only a knowledge of the subject under discussion makes it possible to see the problem if one exists. A good example comes from one of the many books dealing with the intricate issue of tax reform.

Example 10: In *Who Bears the Tax Burden* (Brookings Institute, 1974) by Pechman and Okner (p. 10), the authors explain that the distribution of taxes presents quite a different picture depending on who one assumes ultimately pays corporation and property taxes. "If these taxes are assumed to be taxes on capital, income from capital bears a much heavier tax than income from labor. For example, (under this assumption) the average tax rate on income from capital is 33.0 percent compared with 17.6 percent for income from labor. But the difference is narrowed considerably if the corporation income and property taxes are assumed to be paid in whole or in part by consumers. . . . [(Under this assumption)] income from capital bears an average tax rate of 21.0 percent, while labor income bears a tax of 16.0 percent."

Note that the reason the income tax rate for both capital and labor is lower under the second assumption is that if property and corporation taxes are passed on to the consumer, as rent and price increases, they are no longer income taxes but consumption taxes, not taken into account here. At any rate, we have two different sets of assumptions and there is a considerable difference in relative taxation depending on which we accept. People who live in areas where landlords are permitted by law to pass on 100% of all tax increases directly to the tenant will have no difficulty in deciding which of these assumptions seems more realistic to them.

Irrelevant Percents. Because people tend to see numbers as exact measures, advertisers often throw in a percent or other measure that is really quite irrelevant, just to make it sound as though they were talking about something real or measurable.

Example 11: A recent cigarette ad refers to a study where smokers tested Merit brand cigarettes against others for taste. The ad claims: "The results were conclusive: Even if the cigarette tested had 60% more tar than Merit, a significant majority of all smokers reported new 'Enriched Flavor' MERIT delivered more taste." This 60% was also prominently displayed in the headline.

What does the 60% actually mean? Not very much. All the study actually shows is that Merit was tested against five "low tar" cigarettes and that more than half the smokers preferred Merit. There are literally dozens of brands of cigarettes on the market, ranging in tar levels from 1 mg. to 21 mg. or more. That Merit was able to find five brands which smokers did not like as well as Merit is not terribly surprising. That one of these brands had 60% more tar is irrelevant. We still don't know anything about how Merit compares with all "low tar" cigarettes.

Choosing the Most Flattering Measure. Whether we use percents or the actual numbers on which they are based can depend on what we are trying to hide as well as on what we want to show. When a base is insignificantly small, for example, a percent is preferable to exact figures. If I talk to four friends and three of them plan to vote for Jones in the upcoming election, I can write an article in which "75% of those interviewed expressed a preference for Jones." Alternatively, if a poll shows that out of 5000 people interviewed, 2300 were for Smith and 2700 for Jones, I might prefer to use raw data instead of percents and say something like "Hundreds more of those interviewed planned to vote for Jones than for Smith."

Rank Errors. Ranks and percentile ranks also can be used deceptively. Any product can be shown to rank first if the group in which it is presented is chosen with enough care.

Example 12: A recent ad for Kent cigarettes says, "Here's how the U.S. Government ranks all these cigarette brands. (Compare your brand with Kent Golden Lights.)" It then proceeds to list 36 cigarettes including different sizes and brands according to the amount of tar and nicotine they contain. Kent ranks first, with the lowest amount of each.

The impression that one gets from reading the ad is that Kents rank first of all cigarettes. But of course that is not true. It ranks first only of "all these" cigarettes, the ones selected to appear on the list. Several other brands, advertised in the same newspaper or magazine, have lower tar and nicotine contents, but the ad writer did not include these in the list.

There are many more examples that could be given of the misuse of percents in both the premises and the arguments of advertisers, politicians, business executives, and even scientists. If you learn to question everything you read, you will no doubt come up with many new examples yourself. But you should also be able to have more confidence in honest statements and careful research, which will stand up to the kind of critical judgment you will be exercising.

SUMMARY

PART A

Comparisons can be made either by subtraction or by division. Comparisons by division can take these forms:

a. The division can be indicated without being performed and the numbers expressed as a ratio, either in the form a:b or as the fraction a/b, reduced to lowest terms. If a is larger than b, the division is usually performed, so that if a/b = x, we can say a is x times as big as b or b is 1/x of a.

b. The fraction a/b can be expressed as a percent.

The **base** of a ratio or of a percent can be thought of as the whole amount as opposed to a part, or as the amount that the first quantity is being compared to. In the fraction a/b, the base is the denominator, b. In a percent, if we think of a% as meaning a/100, then the base is the amount represented by the 100.

Rates can be used to measure increases or decreases over a period of time by subtracting the smaller amount from the larger and dividing the result by the amount that is *earlier* in time. Rates are often expressed as percents.

Increases or **decreases** in **rates** are figured in the same way as rates, using the rates as the amounts.

Scores can be expressed as **raw scores** (the number right or wrong) **percent scores** (usually the percent right), **ranks** or **percentile ranks.**

To calculate **rank,** begin with the highest score which has a rank of one, continuing to assign each score a successive number. If there are several instances of the same score, their ranks are averaged.

Percentile rank is figured using the formula:

$$PR = \frac{L + \frac{1}{2}S}{T}$$

where PR = percentile rank, L = the number of lower scores, S = the number of same scores and T = the total number of scores.

PART B

The following errors, committed intentionally or not, result in the misuse of ratios, ranks or percents and often create false impressions or untrue statements.

Poorly defined base   **Irrelevant percents**
Incorrect base   **Misleading percents**
False assumptions   **False comparisons**
                                 Use of most flattering measure

EXERCISES

PART A

Use the following data for exercise #1–#5. The data comes from *The World Almanac & Book of Facts*, 1974.

| | U.S. NATIONAL INCOME IN MILLIONS OF DOLLARS | |
	1971	1972
Wages and Salaries	$573,832	$627,334
Corporate Profits	$85,053	$98,037
Other Sources	$200,053	$216,421
Total National Income	$859,449	$941,792

1.a. *How much more money was earned in wages and salaries in 1972 than in 1971?

   b. *How much more money was earned in corporate profits?

2.a. What was the ratio of wages and salaries earned in 1971 as compared to 1972?

   b. What was the ratio of corporate profits in 1971 compared to 1972?

3.a. *What percent of the total national income was wages and salaries in 1971? In 1972?

   b. *What percent of the total national income was corporate profits in 1971? In 1972?

4.a. What was the rate of increase in wages and salaries from 1971 to 1972?

   b. What was the rate of increase for corporate profits?

5. *Suppose that the rate of increase in wages and salaries from 1972 to 1973 was 10% while the rate of increase for corporate profits was 17%. Was the rate of increase for wages and salaries increasing at a faster or slower rate than the rate of increase for corporate profits? Calculate the increase in both rates and compare.

6. Being able to find rates and percents by doing rough estimates often helps one to understand newspaper reports very quickly and with a

minimum of reading. The following table accompanied a story in the *New York Times* of September 9, 1976 (p. 34). What is interesting here is the relationships between the cuts in budget, enrollment and faculty and staff. See how quickly you can express each of these reductions as a percent by rounding off the relevant figures to one significant digit and estimating the answers. Approximately what percent of the budget was cut? Of the enrollment? Of the faculty and staff?

Cutbacks at City University

Budget (Millions of Dollars)

	1975–76 (actual)	1976–77 (budget)	Difference
State Contribution	$224.0	$174.5	− $50.0
City Contribution	204.3	160.5	− 43.8
Tuition and Fees	70.3	135.5	+ 65.2
Total	$498.6	$470.0	− $28.6

Enrollment (Full-time and Part-time Undergraduates)

	September 1975	September 1976 *	Difference
Senior Colleges	118,340	95,395	− 22,945
Community Colleges	68,994	59,645	− 9,709
Total	187,334	155,040	− 32,294

Faculty and Staff

	September 1975	September 1976 *	Difference
Full-time Teachers	9,642	7,800	− 1,842
Part-time Teachers	5,632	3,000	− 2,632
Nonteaching Professionals	2,348	1,900	− 448
Total	17,622	12,700	− 4,922

* Estimated   *Source: City University*

7. *Below are some test scores from a class of 25 students. These are raw scores from a test with 50 items on it, each score indicating the number of items completed correctly. Change each score to a percent score.

   18, 35, 45, 27, 20, 25, 37, 30, 48, 35, 27, 36, 37, 38, 25, 20, 30, 35, 34, 38, 30, 38, 32, 38, 34.

8. Using the data in Exercise 7, find the rank of the following raw scores: 18, 30, 34 and 38.

9. *Find the percentile rank for each of the scores in #8.

10. What does the percentile rank for the score 38 tell you?

PART B

1. Explain what is wrong with each of the following examples.

   a. *From the *Boston Herald American* of February 10, 1976:

Profit Reports—Major Oil Companies
for Year Ended Dec. 31, 1975

Company	Profit Trend	Profit on Sales	Profit per Gal.
Exxon	Down 18 percent	5.1 percent	2.3 cents
Texaco	Down 46 percent	3.3 percent	1.5 cents
Shell	Down 16 percent	4.3 percent	1.9 cents
Gulf	Down 34 percent	4.3 percent	1.9 cents
Sun Oil	Down 42 percent	5.0 percent	2.2 cents
Union Oil	Down 19 percent	4.5 percent	2.0 cents
Standard Oil (Ca.)	Down 26 percent	3.9 percent	1.8 cents
Cities Service	Down 32 percent	4.2 percent	1.9 cents
Marathon	Down 25 percent	4.0 percent	1.8 cents

   In short, not only are profits down for nearly every oil company, the overall percentage profit on sales is below the national average for all companies, big and small, in all industries.

   b. Quoted from *How to Lie With Statistics:* "The death rate in the Navy during the Spanish-American War was 0.9%. For civilians in New York City during the same period it was 1.6%. Navy recruiters later used these figures to show that it was safer to be in the Navy than out of it."

   c. *Taxes go up by 15%. The landlord uses this as an excuse to raise rents by 15%.

   d. If inflation continues at the rate of 12% a year for five years, an article costing $1 today will cost $1.60 in five years.

   e. *In 1973, about 40% of all men in the labor force and 60% of all women held white collar jobs. From this we can conclude that there were more women white collar workers than men.

2. Suppose you are a member of the class which received the test scores given in problem #7 above. You scored 32. How can you express this score to make it look as good as possible? Defend your answer.

Chapter 3

GRAPHS

PART A: Reading and Constructing Graphs

The purpose of a graph is to allow a quick overview of numerical data so that the important points are emphasized. A graph gives no information that a table couldn't provide, but it makes that information much easier to grasp. If we are concerned about detail, then a table is probably more useful, but a graph is often the best way to show trends, predictions and comparisons. There are many different kinds of graphs, some taking the shape of circles or rectangles, others of pictures or maps. We will begin with circle graphs.

Circle Graphs

These are in some ways the simplest graphs because each graph only deals with one set of data. That is, a circle can only show how something is divided up; it cannot make comparisons between two or more sets of things. In order to make a comparison, we must use two or more circle graphs, as in Example 1 below. Circle graphs are often used to show how a budget divides up the total income or total expenditures for the year. Or they can be used to show the makeup of a particular population, as in this example from the *New York Times* of Sunday, Feb. 23, 1975.

Example 1:

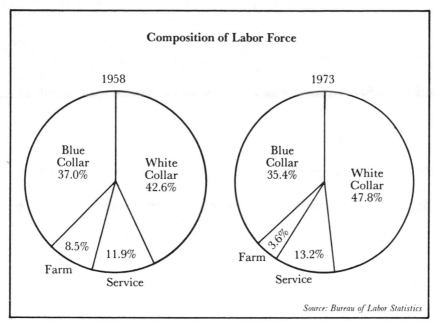

Composition of Labor Force

The circle graph is very easy to read. The title (Composition of Labor Force) tells us what the circles represent. Each section is labeled with a percent that tells us numerically what part of the whole it represents. The relative sizes of the sections give us a visual impression of what part each section is of the whole. For example, in 1958 farmers represented 8.5% of the labor force and by 1973 they represented only 3.6%. We can tell at a glance how things have changed by comparing the two circles.

Rectangular Graphs

These are used when we wish to show the relationship between two sets of data in one graph. It is not necessary for both sets to be numerical, though one always is. The other set might represent the months of the year, the names of different countries, or a list of products, or it too may be numerical.

Suppose we want to make a graph to show how corporate profits have risen since 1950. Using the 1977 *Statistical Abstract of the United States*, we find the following data (Table 636):

Year:	1950	1955	1960	1965	1970	1975
Profits (in billions):	$33.7	$44.6	$46.6	$77.1	$67.9	$91.6

A rectangular graph shows the relationship between two sets of data by using a scale to measure one set along its base, or horizontal axis, and a scale for the other along its height or vertical axis. It doesn't really matter which scale goes on which axis, although dates often go along the horizontal one. Any scale must be evenly spaced so that equal differences between numbers take up equal space. In our graph, which uses every fifth year, the dates will be evenly spaced along the horizontal axis. If the dates were 1950, 1960 and 1965, we would leave twice as much space between the first two as between the second and third dates. The vertical scale must be marked off in even units as well. It is not always necessary to start at zero, as we shall see later, but it does make a good reference point and we shall use it here. We can then work up in any convenient intervals (let's use $10 billion) to a number above the highest we will use ($100 billion makes a good stopping place in this case). It is important to clearly label the scales so that there can be no mistake about what units we are using. Finally the graph needs a title that describes exactly what its subject is. What we have so far looks like Example 2a.

We are now ready to turn our data into points on a graph. Our first point comes at the intersection of the year 1950 and the amount $33.7. That is, we start at the base of the graph at 1950 and move up to the point where we are directly opposite $33.7 (or the point where we estimate $33.7 to be on the scale). See Example 2b. Similarly, the next point is exactly above 1955, and to the right of $44.6. We continue until we have a dot for each pair of numbers. Our completed **dot graph** is shown in Example 2c.

Example 2:

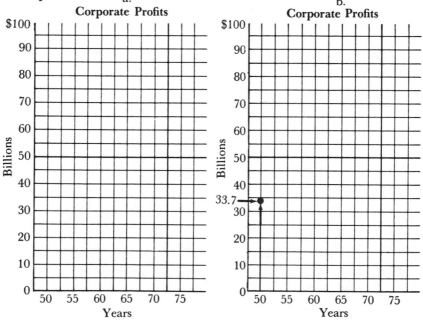

c.
Corporate Profits

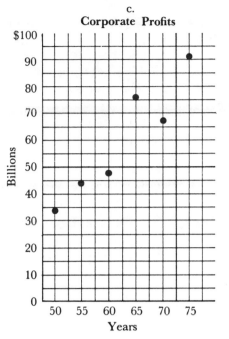

Not a very dramatic graph, is it? One way to make the picture clearer is to connect all the points and make a **line graph,** like this:

Example 3:

Corporate Profits

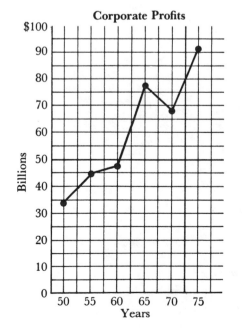

This type of graph is frequently used to indicate trends. In our present case, though, it is not altogether appropriate. The line seems to show that there was a smooth rise from 1950 to 1955. But we don't know from the data that this is true. There may have been a severe slump, say in 1952, with profits going way down and then up again, for all we know. If the data for 1970 had been omitted, we'd think there had been a continuous rise from 1965 to 1975. So although this form of graph is very useful when there is enough information to justify it, we need to present our data on corporate profits in a different form.

Another way to show changes is to use a **bar graph.** Since it is made up of separate bars, there is no danger of indicating a false continuity. Example 4 presents the same data on corporate profits in bar graph form.

Example 4:

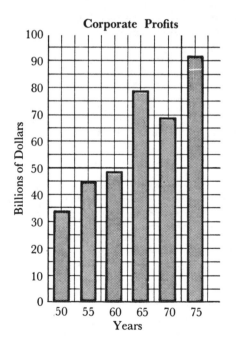

Bar graphs are also useful for presenting data which is naturally disconnected, such as the total amounts of iron, steel, and copper produced in a year. Bar graphs can be either vertical or horizontal.

Composite Graphs

Dot, line or bar graphs can also be used to show the interaction of more than two factors. You may wonder how this is possible when a rectangular graph has only two dimensions, and so only room for two scales. The secret is to combine two or more graphs to make a composite graph.

Example 5:

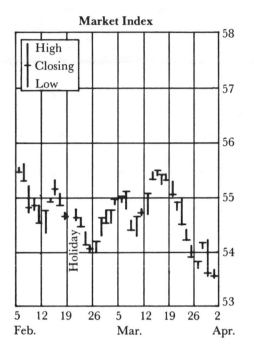

Market Index

Example 5, from the *New York Times* of April 1, 1977, shows a dot graph which plots against each weekday the high, low, and closing prices on the New York Stock Exchange. We can see, for example, that on April 1, the highest index price reached during the day was about 54.2, the lowest about 53.6, and the closing price was just a shade above the lowest.

Exercise: What were the high, low and closing prices on Feb. 8?

Graphs like this are used not only to show the activity of the stock market, but also in scientific work where it is necessary to record high, low, and average values over a period of time.

There are two kinds of **composite line graphs,** as we can see from Examples 6 and 7. Example 6 is from *Dollars and Sense,* No. 21, November, 1976; Example 7 is from the *Boston Globe* of January 21, 1976.

Here we have two separate graphs placed on top of each other. Note that each uses the same base line and each has a separate vertical scale. We can see, for example, that in 1976 the unemployment rate for the whole working population was 8% and women represented 40% of the labor force.

Example 6:

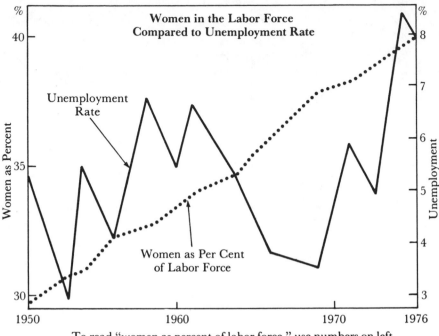

Women in the Labor Force
Compared to Unemployment Rate

To read "women as percent of labor force," use numbers on left
edge of graph. To read "unemployment rate," use right edge.

This type of graph is useful for making comparisons, but it has limitations. One must identify each line with its title and keep it distinct. In most cases more than 3 or 4 categories would lead to a very crowded and confusing graph.

In Example 7 we have separate graphs put together, but here a different principle operates. The five graphs are piled on top of each other in such a way that the top of one graph area forms the base line for the next. In order to see how much was spent in each category, we subtract the amount represented by the bottom line from the amount represented by the top line. Suppose we want to know how much was spent for R & D (weapons research and development) during 1970. We look first at the intersection of 1970 and the lower border of the R & D band. From the vertical scale we can see that it represents about $48 billion. Looking next at the upper border of the R & D band, we estimate a value of about $55 billion. Subtracting the first from the second value, we find that the 1970 expenditure for R & D was about $7 billion.

Exercise: Estimate the amount spent on Retired Military Pay in 1975.

Example 7:

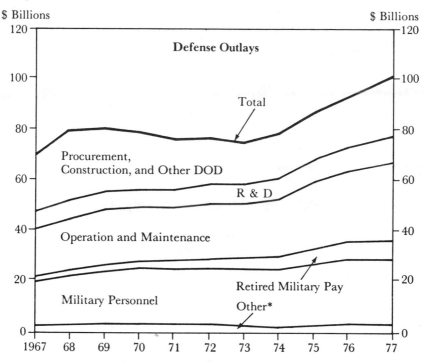

$ Billions $ Billions

Defense Outlays

Total

Procurement,
Construction, and Other DOD

R & D

Operation and Maintenance

Retired Military Pay

Military Personnel   Other*

1967   68   69   70   71   72   73   74   75   76   77

On the one hand, this kind of graph is not particularly useful if we are interested in details such as exact expenditures in a particular category for a given year. On the other hand, it does serve a useful purpose in showing us trends over a period of time. We see for example that the expenditure for R & D has remained fairly constant over the years, that the amount required for Retired Military Pay has increased considerably and that with one exception there are no significant decreases in expenditure over time.

Exercise: In which category does the exception occur?

The same techniques that are used with line graphs can be applied to bar graphs to form **composite bar graphs.** Look at the next example, from *Fortune* of January, 1977.

Example 8:

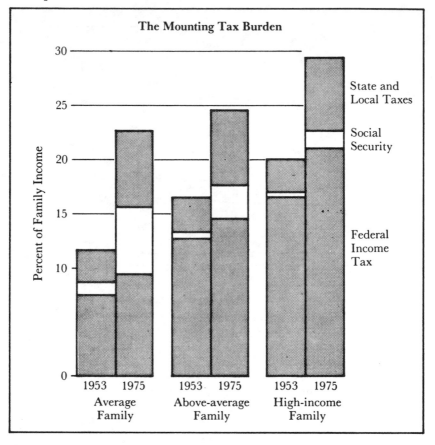

Like the graph in Example 7, this one uses the technique of piling several graphs on top of each other, with the top of the first graph (federal income tax) providing the base line for the second (social security) whose top is the base for the third (state and local taxes). Notice that the total of all three amounts in this graph (and of all six amounts in Example 7) can be read directly off the scale, since in both cases the top line represents the total of all the categories. In the graph in Example 7, the top line is labeled "Total"; in Example 8 it is assumed that the top of each column represents the total tax burden.

Exercise: Estimate the percent of its income an average family spent on social security in 1975.

Example 9:

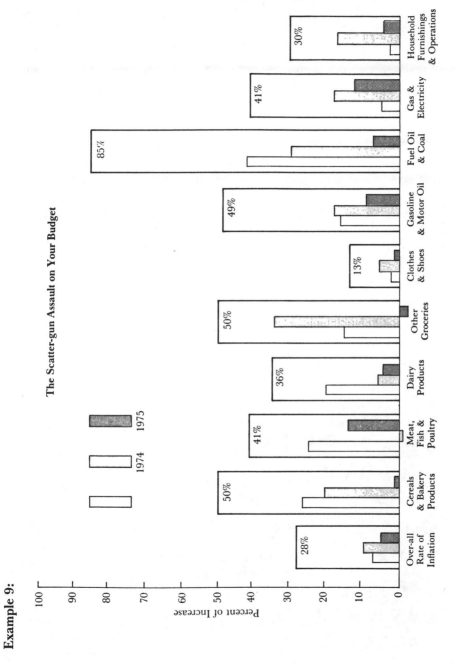

The Scatter-gun Assault on Your Budget

Our final composite graph from *Changing Times* of May 1, 1976, is really a double composite. It shows price increases for three years by bars placed next to each other using the same base line (the three thin columns), but it also shows the amounts when the three individual columns for each category are combined (the superimposed wide columns).

Exercise: To test your understanding of this graph, estimate the increase in the cost of gas and electricity in 1974. What was the increase in gas and electricity for all three years, 1973, 1974 and 1975 combined?

Picture Graphs

Graphs using pictures are often more eye-catching and less intimidating than the more technical graphs we have been discussing. Basically there are two types, those in which the quantity is represented by the number of objects, as in Example 10, and those in which the quantity is represented by the size of the objects. We will discuss the latter type in the next section. Example 10 comes from the *New York Times* of July 4, 1976. From this graph we can figure out what percent of the population over 16 years of age works for local, state, and federal government. The graph tells us that 9 out of every 100 people are government employees and also represents these by 9 stars, so we know that 9% works for government.

Example 10:

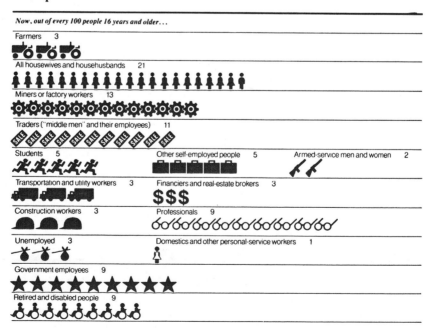

Now, out of every 100 people 16 years and older...

Farmers   3

All housewives and househusbands   21

Miners or factory workers   13

Traders ("middle men" and their employees)   11

Students   5   Other self-employed people   5   Armed-service men and women   2

Transportation and utility workers   3   Financiers and real-estate brokers   3

Construction workers   3   Professionals   9

Unemployed   3   Domestics and other personal-service workers   1

Government employees   9

Retired and disabled people   9

Another form of picture graph is the **geographic graph.** This presents information about geographic areas by superimposing the relevant data on a map of the area in question. The following example is taken from an article in *Dollars and Sense,* No. 46, April 1979, about manufacturing employment and plant closings in the U.S. The following graph shows the percent and the actual number of manufacturing jobs lost or gained from each part of the country.

Example 11:

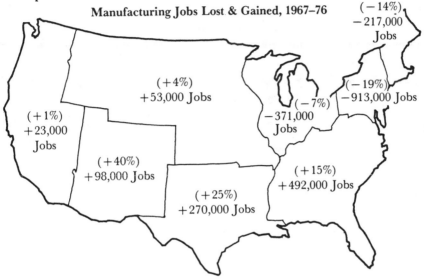

Manufacturing Jobs Lost & Gained, 1967–76

There are many more kinds of graphs than we have been able to show here. However, all the basic principles are here. When we encounter a new type of graph, the best approach is to consider in turn the following questions:

1. What is the graph referring to — what is its title?
2. What units are being used — what are the scales?
3. How are the quantities being represented, by an area (as in a circle or picture graph) or by a point, as in a rectangular graph?
4. If the graph is composite, what is the base level for each category?

PART B: Graphical Illusions

The main advantage of graphs is that they can give us a quick over-view of a situation. But this advantage can also be a drawback. If a graph can give a true impression, it can also give a very false impression just as quickly. When we look at a graph, we are less concerned with the details of actual numbers and more with the general picture of the situation it gives us. We note how steeply the line slopes, or how much of the area is colored in. We are less likely to be concerned about the scales that give meaning to those visual impressions.

Let's look at a particular example. Suppose we want to graph black income as a percentage of white income over the last 25 years. The data is given in the table below.

Year:	1950	1955	1960	1965	1970	1975
Percent:	54%	55%	55%	55%	64%	61%

The first step in setting up a graph is to choose the scales. Here we might run the vertical scale from 0 to 100% and the horizontal from 1950 to 1975. See Example 12.

Example 12:

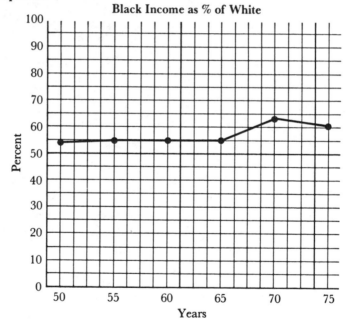

Black Income as % of White

Notice that the impression the graph gives is that after 15 years of a stable state of inequality, black income improved from 55% of white income to 64%, and then began to decline again. Suppose, now, that we wanted to make the picture look less dismal or alternatively, even more dismal. We can use two techniques: cropping and distortion, either of which will dramatically change the picture.

Cropped Graphs

The graph in Example 12 has a lot of waste space on it. Since the line never goes below 50% or above 70%, we could reduce the picture to that area, as in Example 13. Notice that doing away with the blank spaces emphasizes the rise in the graph and makes the change look more significant than it did in Example 12.

Example 13:

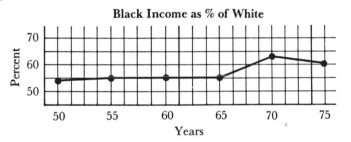

Besides cropping to emphasize change, we can also crop to make the general situation look better or worse. Look at Example 14A and 14B.

Example 14:

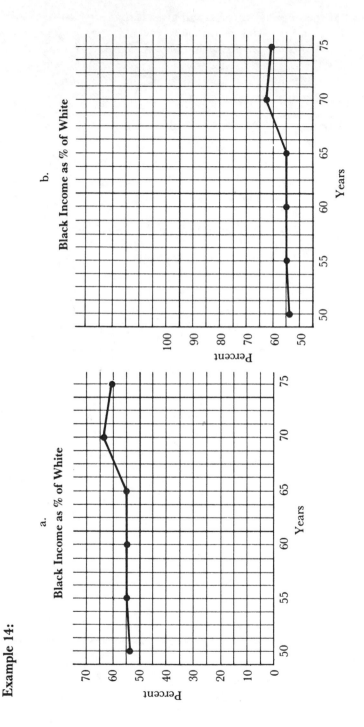

a.

Black Income as % of White

b.

Black Income as % of White

Example 14A makes blacks' income look "higher" than example 14B.

We can also select only that part of the data which we want to show by cropping the horizontal axis. If we want to present a stable followed by an improving situation, we can forget about 1975. If we want to show a volatile situation, we start with 1965. Horizontal cropping of this kind is simply a form of suppressed evidence, a fallacy we discussed in Chapter 1.

Suppose now that we want to present our data as fairly and clearly as possible. The graph as we presented it in Example 12 does not mislead, but neither is it an attractive graph. There is a great deal of unused space. Must we always start with zero on our scale? The answer is no, it is not always necessary. But it is important to call attention to what we are doing. Readers are likely to assume that the scale starts at zero, so it is only fair to state explicitly when it does not. The technique used by graph makers, as shown in Example 15, is to use a broken line in the scale to indicate where it has been cropped. The actual graph can then be centered in the area indicated to give a balanced picture that does not misrepresent the data.

Example 15:

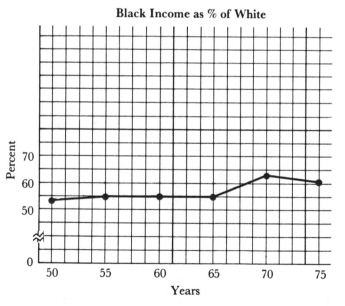

Distorted Graphs

An even more effective technique for manipulating the impression given by a graph is to vary the scales. Since the horizontal and vertical units on a graph rarely represent the same units, there is no built in relationship between the two. For example, if 10% on the vertical scale equals half an inch, how much space should be given to 10 years on the horizontal scale? There's no rule about it — how could there be? But it makes a great deal of difference what we decide the relative distances will be. On the one hand, proportionately larger distances along the vertical scale tend to emphasize change (Example 16A). On the other hand, larger distances along the horizontal scale emphasize stability (Example 16B).

Example 16:

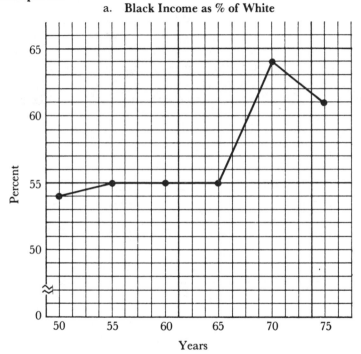

a. **Black Income as % of White**

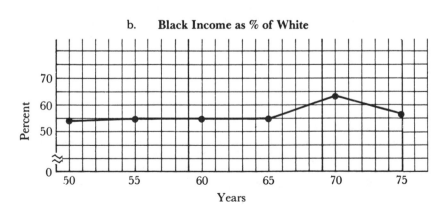

b. **Black Income as % of White**

If you wanted to show that unemployment had increased a great deal in the last few years, would you put the dates on the horizontal axis farther apart or closer together? Would you put the percent of unemployment on the vertical scale farther apart or closer together? Notice that either putting the dates close together, or putting the percents far apart would serve your purpose. Doing both simultaneously would further increase the impression you are trying to make.

False Comparisons

Although there are no set rules beyond common sense to determine the relationship between the horizontal and vertical scales of a particular graph, the case is different when we compare two separate graphs. If both are dealing with the same subject, then it is expected that the vertical scale of one will match the vertical scale of the other. The same should be true for the horizontal scales. In Example 17 we can see what happens when this rule is not followed. These two graphs appeared in *Newsweek* on May 9, 1977. Because the "total major costs" and the "government expenditures" lines look so much alike, a casual observer could not help but come to the conclusion that most medical costs are met by the government. Yet when you look at the scales, you see that the top graph ranges from 0 to 100 billion dollars, while the bottom one reaches only 50 billion.

Example 17:

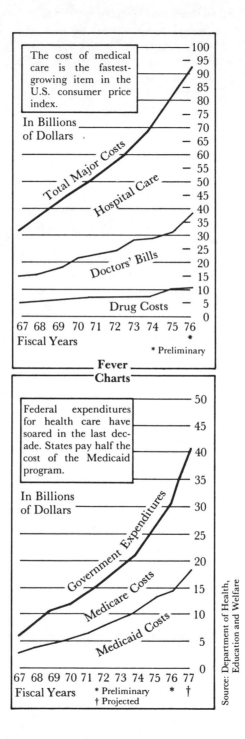

The cost of medical care is the fastest-growing item in the U.S. consumer price index.

In Billions of Dollars

Total Major Costs

Hospital Care

Doctors' Bills

Drug Costs

67 68 69 70 71 72 73 74 75 76

Fiscal Years

* Preliminary

Fever Charts

Federal expenditures for health care have soared in the last decade. States pay half the cost of the Medicaid program.

In Billions of Dollars

Government Expenditures

Medicare Costs

Medicaid Costs

67 68 69 70 71 72 73 74 75 76 77

Fiscal Years   * Preliminary   * †
                                    † Projected

Source: Department of Health, Education and Welfare

The techniques of cropping and scale distortion can be used with any type of rectangular graph. There is another graphical fallacy which is peculiar to picture graphs.

Area Graphs

As we mentioned in Part A, there are two kinds of picture graphs, one of which we showed in Example 10. That graph represented each quantity by the appropriate number of objects. The other method represents the quantity by the size of the object, as in Example 18 from the July 4, 1976 issue of the *New York Times*.

Example 18:

The women march out of the kitchen

In 1776, only about five percent of white women were in the labor force. Now that figure is approaching 50 percent. As the chart shows, black women have always worked outside the home in great numbers. The trend toward work is most pronounced among mothers of children under six—in 1975, 39 percent of such women were working, as opposed to only 14 percent in 1950. Is the absence of the mother harmful to little kids?

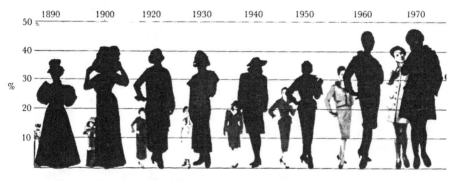

Picture graphs which depend on the height of the object, as measured against a vertical scale to represent a quantity, always give the wrong impression. The problem is that the overall impression when we look at the graph is due to the area of the figures, and not to their height. Look at the figures for the year 1930, which are supposed to show that 40% of black women and 20% of white women worked outside the home. The figure for black women is twice as tall, but it looks four times as *big;* and it is four times as big in area. Bar graphs do not present this problem, because bars can be made taller without becoming wider. To make a human figure, or any other picture, taller without making it wider means stretching it into an unnatural shape.

Propaganda Graphs

So far we have been looking at graphs which contain errors, or poor techniques, that may or may not be intentional. There are some fallacies, however, which cannot be given the benefit of the doubt, and whether they are used to convince the reader of a political point or to sell a product, the result is a blatant distortion of the data. One of the simplest techniques in this area is to present a graph with presumably accurate data, but to label it in such a way as to give a totally false impression. The graph below is from *Fortune,* October 1976.

Example 19:

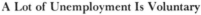

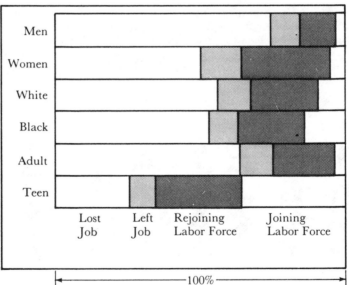

Job losers are barely over half of all the unemployed, and are mostly adults. Many others are unemployed because of actions they took more or less voluntarily: they quit their last job, just joined the labor force, or just rejoined it. Overall unemployment would be substantially reduced if these newcomers were able to find jobs more quickly.

It is difficult to see by what stretch of the imagination reaching the age of employment and being unable to find a job qualifies as voluntary unemployment!

Another method for giving false impressions is to simply leave out one or more scales when constructing the graph. Then one can have lines running up or down the page at whatever angles one wishes without danger of being held accountable. Add to that graphing the future rather than the past, and you have a graph like this one, run by the American Gas Association in many leading magazines during 1976.

Example 20:

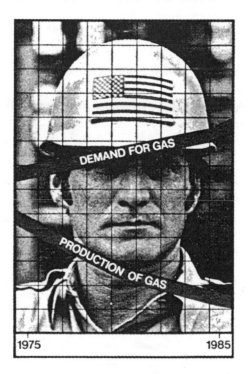

The previous graph also adds another feature which we have not yet discussed, and that is the use of pictures or other symbols within the design of the graph which do not provide additional information, but are meant to have an emotional impact. In this example, we see that the patriotic American worker looks grimly on as the decreased production of natural gas costs him his job. The text spells out the message, the picture gives it impact, but where are the facts?

SUMMARY

PART A

The following kinds of graphs are in common use:

a. Circle graphs: quantities are represented by sections of a circle
b. Rectangular graphs
    i. Dot graphs: two pieces of information represented by a single dot
    ii. Line graphs: dots are connected by lines
    iii. Bar graphs: solid bars from dots to base
    iv. Composite graphs: can be dot, line or bar, representing several sets of data

c. Picture graphs
    i. Quantity represented by number of objects
    ii. Quantity represented by size of objects
    iii. Geographic graphs

PART B

False impressions can be conveyed graphically through the following methods:

a. Cropped graphs: cropping top gives impression of higher values, cropping bottom gives impression of lower values, cropping both emphasizes fluctuations.
b. Distorted graphs: stretching vertical (quantity) axis emphasizes change, stretching horizontal (time) axis emphasizes stability.

c. False comparisons: compared graphs of same phenomenon should have same scales.

d. Picture graphs: two-dimensional objects of different heights have disproportionately different areas.

e. Purposely misleading graphs: The following characteristics are common in graphs meant to deceive:

    i. Titles which contradict data of graph

    ii. No scales

    iii. No discernible data on which graph is based

    iv. Pictures or symbols designed to arouse emotion instead of conveying information.

EXERCISES

PART A

1. The graph below, loosely modeled on the much more detailed chart by Stephen J. Rose (*Social Stratification in the United States,* Social Graphics Co., Baltimore, 1979) represents the income and wealth of the population of the United States. This is an example of two separate graphs in one frame. The graph on the left represents wealth. Each dollar sign represents $100,000 of wealth owned by each person at the percentile rank found on the left hand vertical scale. The graph on the right represents income. Each figure represents about 3% of the population. The vertical scale on the right represents annual income in thousands of dollars.

    a. What is the amount of wealth held by the average person in the upper .2% of the population? If the population is in the neighborhood of 200 million, how many people possess that much wealth?

    b. What percent of the population earns less than $5,000 a year? How many people does that represent?

Source: Stephen J. Rose, *Social Stratification in the United States: An Analytic Guidebook* (Baltimore, Md, 1979) pp. 5, 28

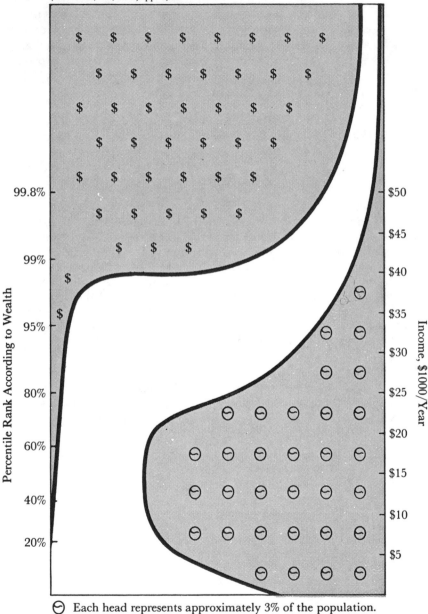

⊖ Each head represents approximately 3% of the population.

$ Each dollar sign represents $100,000 of wealth owned by *each* person at that percentile rank.

2. Sketch a circle graph to represent the information in Example 10. You may want to group some categories together. Label each section carefully.

3. *Make a line graph to illustrate the following data:

Unemployment Rate for Blacks and Other Non-White Races 16 Years Old and Over

1955	1956	1957	1958	1959	1960	1961	1962	1963	1964	1965
8.7	8.3	7.9	12.6	10.7	10.2	12.4	10.9	10.8	9.6	8.1

4. Make a bar graph to show the following:

Purchasing power of the dollar: 1967–1974
(measured in 1967 dollars)

1967	1969	1971	1973	1974
$1.00	$.92	$.82	$.73	$.66

5. *Use whichever type of graph you consider most appropriate to display the following information; many different solutions are possible — none of them perfect.

WOMEN IN THE LABOR FORCE

Occupation	Black females		White females	
	1950	1974	1950	1974
White Collar	13%	42%	60%	63%
Blue Collar	19%	20%	24%	15%
Service Workers	67%	36%	15%	20%
Farm	1%	2%	1%	2%

6. List at least five conclusions that can be drawn from looking at your graph in #5.

PART B

1. What criticisms can you make of the graphs on pp. 74–75?

a. *From *Newsweek*
    June 6, 1977

A Gap in Interest

Percent Interest Rates

Prime Rate

| | 1974 | 1975 | 1976 | 1977 |

Source: Federal Reserve Board

b. From the *Boston Globe*
   July 1, 1977

% of Calories from Fat

* U.S. Diet:

Current
42%

Goal
30%

Milk:

Whole
48%

Nuform
20%

* U.S. Senate Committee Report "Dietary
Goals for the United States" Page 12

c. * From *Newsweek,*
    June 10, 1974

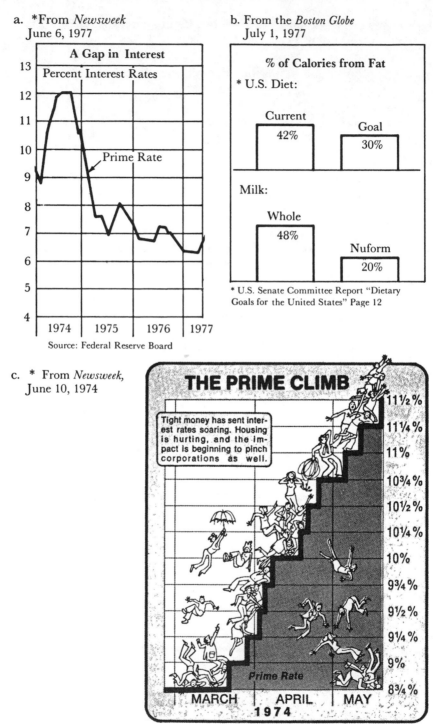

THE PRIME CLIMB

Tight money has sent interest rates soaring. Housing
is hurting, and the im-
pact is beginning to pinch
corporations as well.

11½ %
11¼ %
11 %
10¾ %
10½ %
10¼ %
10 %
9¾ %
9½ %
9¼ %
9 %
8¾ %

Prime Rate

MARCH   APRIL   MAY
1974

Drawing by John Huehnergarth; chart by Fenga & Freyer

d. From *Business Week,*
February 2, 1974

e. * From the *New York Times*
March 22, 1977

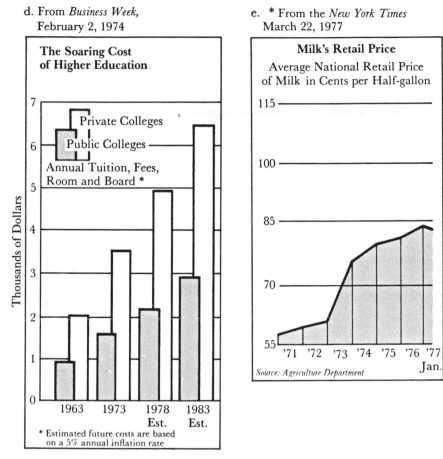

f. From the *New York Times,* July 4, 1976

The birth trend is down, down, down . . .

Total fertility rate: average number of children per woman of childbearing age.

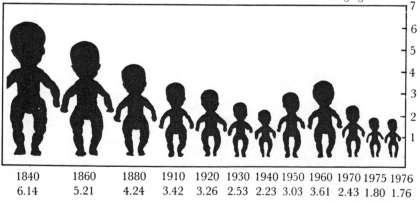

	1840	1860	1880	1910	1920	1930	1940	1950	1960	1970	1975	1976
	6.14	5.21	4.24	3.42	3.26	2.53	2.23	3.03	3.61	2.43	1.80	1.76

2. Construct two graphs for the following data, one from the point of view of the union, the other from the point of view of management.

Hourly Wages of Widget Makers, 1969–1974

1969	1970	1971	1972	1973	1974
$2.25	$2.35	$2.50	$2.80	$2.95	$3.25

3. Look through some newspapers and magazines and see if you can come up with any examples of graphical editorializing.

Chapter 4

INDICATORS
AND INDEXES

PART A: The Magic Numbers

Economic Indicators

We saw in Chapter 3 that large amounts of data can be summarized in a picture or graph. In this chapter we will see that it is also possible to characterize a set of data with a single number. If you have ever looked at the economic pages of a newspaper, you have encountered several single number summaries called economic indicators. Additionally, the word "index" is sometimes loosely used to refer to economic indicators. Some of these use averaging techniques; others do not. Perhaps the most familiar and often quoted of these indicators is the GNP (Gross National Product), the figure used to indicate the economic health of the nation by looking at its total output. This number is a sum — the total value of all goods and services produced in a year. Other important indicators are the Dow-Jones Industrial Average, the Consumer Price Index and the unemployment rate. In this chapter, we will look at these four indicators as examples of single number measures of complex phenomena.

The Gross National Product. The measure of Gross National Product, or GNP, is used in the U.S. as a basic indicator of economic activity. Its course over a period of years is taken as a measure of economic health.

There are four major categories considered in calculating the GNP: personal consumption, private investment, the net difference between exports and imports, and government purchases of goods and services. In each category the current price of each item is multiplied by the number of items of that kind produced. Then the total value of each type of item in each category is added up. Finally the values of the four categories are added to arrive at the GNP.

It is important here to note that what is being measured is goods and services produced for final use during the course of a year. The term "final use" means, for example, that the cloth purchased by dress manufacturers is not directly counted in GNP — it is an "intermediate good" whose price is included in the price of dresses sold to consumers, so to count it directly would be to count it twice.

Finally, the investment part of GNP needs a little clarification. All investment goods (buildings and machinery) produced in a given year are included in GNP, even though some of them simply replace other buildings and machines worn out in the course of the year's production. Another measure, which will not be discussed here, excludes this replacement investment. (It is called the Net National Product and it subtracts from GNP the value of the worn out buildings and machinery.)

As the most inclusive single indicator of the overall level of economic activity, GNP provides useful information about short-term changes in the state of the economy. Yet it is important to realize that by itself it is not particularly useful. One problem that plagues GNP is that higher selling prices can cause a rise in the GNP which does not reflect any real increase in production. Therefore, when we make comparisons involving money (wages, prices, taxes, etc.) over a period of time, we must take inflation into account in some way. In the case of the GNP, economists use something called the **GNP implicit price deflator,** the ratio of the GNP in current dollars to the GNP in some **base year. Current dollars,** sometimes called nominal dollars, represent the actual price paid for any good or service. **Constant dollars,** sometimes called real dollars, represent the price of that good or service as measured in its price in some given year, called the **base year.** For the GNP figures, the government now uses 1972 as the base year. However, for our purposes of comparison, any year can be used as the base year. Dividing GNP in current dollars by the implicit price deflator results in a figure called constant dollar, or real, GNP. By comparing real GNP from two different years, it is possible to avoid the problem of changing price levels over time. In addition, by using the same base period over the length of time one is interested in, one can calculate both the rate of inflation and the real increase (or decrease) in production.

Exercise: If the GNP for 1977 was $1889.6 billion in current dollars and $1337.3 billion in constant (1972) dollars, what is the GNP deflator for 1977?

The graph in Example 1 below shows the relationship between the GNP in current dollars, the GNP in constant dollars (the base period is 1972) and the economic growth rate (the percent of increase or decrease in actual production from year to year). Study the graph carefully. Notice that this is a composite graph with two line graphs and one bar graph all superimposed on each other. The left vertical scale refers to the two line graphs and is in billions of dollars. The right vertical scale refers to the bar graph and is in percents. Declines in output are reflected both in the down-turn of the line graph showing GNP in constant dollars, and in the bar graph. The line graph shows the amount of decline in dollars. The bar graph measures the decline as percent or rate of decline from the previous year.

Exercise: In what years did real output decline? Looking at the line graph for GNP in constant 1972 dollars, estimate the decline in dollars from 1973 to 1974. Now looking at the bar graph, estimate the rate of decline from 1973 to 1974.

Example 1:

Gross National Product (GNP) in Current and Constant 1972 Dollars: 1960 to 1975

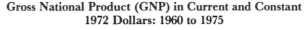

(See Tables 630 and 632)

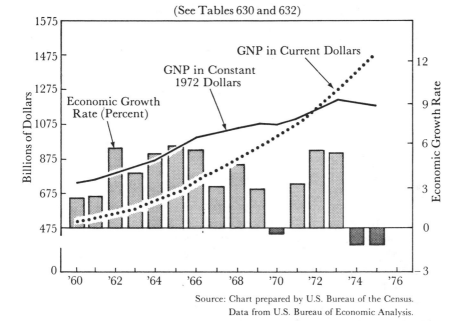

Source: Chart prepared by U.S. Bureau of the Census.
Data from U.S. Bureau of Economic Analysis.

The Dow-Jones Industrial Average. The Dow-Jones Industrial Average is a measure of what is happening in the stock market, where shares in major corporations are bought and sold. The prices that are paid for shares, as they change hands from day to day, are an indication of investors' perceptions of the health of the economy, and the Dow-Jones Industrial Average is one measure of these prices. Other measures are the New York Stock Exchange Index and Standard & Poor's 500. We will look at the Dow-Jones because it is the oldest such measure and the one most frequently reported by newspapers, radio, and TV.

As its name implies, the Dow-Jones is an average of the selling prices of stocks of industrial corporations. (Examples of non-industrial stock are transportation, rails and utilities.) It is not a simple average and it does not include all industrial stocks. Rather, there is a list of 30 stocks, expected to be representative of the market as a whole. Here is the list of stocks on which the average is currently based:

Allied Chemical	General Foods	Owens-Illinois
Aluminum Co	General Motors	Procter & Gamble
Amer Brands	Goodyear	Sears Roebuck
Amer Can	Inco	Std Oil of Calif
Amer Tel & Tel	IBM	Texaco
Bethlehem Steel	Inter Harvester	Union Carbide
Du Pont	Inter Paper	United Technologies
Eastman Kodak	Johns-Manville	US Steel
Exxon	Merck	Westinghouse El
General Electric	Minnesota M&M	Woolworth

These stocks represent somewhere between 25% and 30% of the total volume of stocks traded on the stock exchange in any particular period of time. The daily Dow Jones is calculated by adding up the prices at which each of these stocks sold at the end of the day and then dividing this sum by 1.465.

Over the years, the list of stocks has changed (only two of the original 12 on which the index was based in 1896 remain) in an effort to keep the list representative. The number by which the sum is divided has also changed, from the original figure of 30, to 3 and more recently to 1.465. These changes were made to take into account what happens when a stock splits. For the moment, though, we will not worry about the actual value of the number. As with other economic indicators, the actual Dow-Jones Average on any one day is of very little importance. What matters is its behavior over a period of time. The Dow-Jones tends to fluctuate widely, not only from year to year but from week to week and even day to day. The graph in Example 2 shows the activity of the Dow-Jones during the period from January 1970 to August 1974. The line represents the closing average for each month during that period.

These fluctuations are a reflection of predictions about the market made by people who trade in stocks. The more confidence they have that business will be good, the more likely they are to buy shares. This is true for individual industries as well as for business as a whole. The more people who want to buy a particular share, the more its price goes up. Similarly, the less confidence people have in a particular company or in the stock market as a whole, the more likely they are to sell their shares and

The Dow Jones Industrial Average

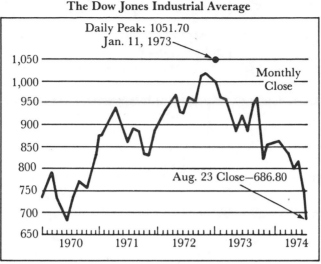

Daily Peak: 1051.70
Jan. 11, 1973

Monthly Close

Aug. 23 Close—686.80

The New York Times/Aug. 24, 1974

cause stock prices to drop. But just what it is that causes any particular increase or decrease in confidence is usually a total mystery to most of us, including many of the so-called experts.

Exercise: What were the highest and lowest values of the Dow-Jones Index in the period from January 1971 to July 1973 in Example 2?

Up to this point we have looked at several indicators such as the GNP, the GNP deflator, and the Dow-Jones Industrial Average. In this section we will look at one important index: the Consumer Price Index. It is the one which is probably most important to us in our daily lives. The Consumer Price Index, or CPI for short, is a measure of changes in consumer prices. Let's see how it is constructed.

The CPI. If we consider for a moment the number of different ways in which a given consumer spends money, the different rates at which items increase in price, the variety of spending patterns among different consumers or in different parts of the country, we see that measuring the change in consumer prices is an enormous undertaking. If we want a measure that will tell each consumer exactly how much his cost of living has increased, the task is impossible. A very general sort of average is the best we can do.

In statistics, we use many different measures that we call averages, a subject we shall treat in more detail in Chapter 6. For now, though, we can take the word average to mean what it usually does in arithmetic: the sum of a set of items divided by the number of items in the set. For example, suppose we are interested in finding the average price of a half-gallon of milk over a year's time. The first step would be to record the price paid for milk every week for 52 weeks. Then we would add up the prices and divide the total by 52, the number of prices we had recorded. The result would be a simple average of milk prices for the year.

To calculate the CPI, the government uses what is called a weighted average. A **weighted average** takes into account the relative importance of a number of different items that are combined in the average. To continue the example of the average price of milk, assume that we recorded prices for two years, 1977 and 1978. We found that the average was $.99 for 1977 and $1.05 for 1978. Now suppose that we are interested in comparing the effect on the consumer's cost of living of the change in the price of milk as compared to a change in the price of detergent. We find that the average price of detergent changed from $.99 to $1.05 between 1977 and 1978. Now look at the figures below.

	1977 average	1978 average	1977 Quantity Purchased remains unchanged	1978
½ gallon of milk	$.99	$1.05	3 times a week	
1 box of detergent	$.99	$1.05	1 every four weeks	

A six cent increase in the price of detergent doesn't have very much impact because we only buy detergent once a month. However, a six cent increase in the price of milk would have a greater impact because we buy milk 3 times a week (or 12 times a month). So in comparing the importance of price increases of milk and detergent it would be inappropriate to only look at the average price of detergent and milk by themselves because we buy more milk than detergent during a year. We would need to calculate a weighted average which would take into account the relative importance of milk as compared to detergent. Detergent would be weighted less than milk because even if its price increases at the same rate as the price of milk, we buy milk 12 times as often.

In Example 7 of Chapter 2, we showed how an overall price increase in a particular commodity (in that example, student's meals) had to take into account the proportional increases of different components of the commodity. We were really using the technique of weighted averaging to find a fair price increase for meals. The CPI can be thought of as a weighted average of price increases in all areas. This kind of weighting is only one aspect of figuring the CPI. There are other problems involved in calculating this index which we can best appreciate by going over some of the steps taken in constructing it.

1. Since CPI measures a change in prices, it must measure that change with respect to some fixed standard. For the CPI, 1967 has been used as the base year since 1971.

2. The CPI is supposed to reflect changes in prices paid by consumers. Since it is impossible to survey everyone, a sample of consumers is selected. The Bureau of Labor Statistics uses a sample that reflects the expenditure patterns of all urban consumers. This index is called the CPI-U (for urban). The CPI-U includes urban wage earners and clerical workers, managerial and technical workers, professionals, the self-employed, short-term workers, the unemployed, and retired people. The CPI-U represents 80% of the population.

3. Extensive research was done in 1972–73 for the CPI-U to determine how the sampled consumers spent their money. Expenditures were first broken down into major categories such as Food, Housing, and Transportation. Then these major categories were subdivided so that Food, for example, was divided into Beef, Pork, etc. After each category was broken down, there were about 400 items in the index. These 400 items make up what is called the **market basket** for the base period.

4. Further research determined what portion of their total income consumers spent on each item in the market basket. That portion of the total income is expressed as a percent or decimal and is used as a weight in the average. For example, if the average family spent 1.47% of its income on milk, then the weight for milk in the index is .0147. All the weights together must add up to 100% or 1.00. Once established, neither the basic market basket nor the weights change between major revisions.

5. The next step was to find out how much the consumer would have paid for the 400 representative items in the 1972–73 market basket at 1967 prices (the CPI-U still uses 1967 as the base year). Then the cost of the 1967 market basket could be calculated, using appropriate prices and weights. The actual procedure used involved several further complications, since weights were also applied for 85 representative communities.

6. On a regular basis surveys find the current prices. For some items in the market basket, current prices are checked every month, for others every three months.

7. Each month, the current CPI-U is calculated. Although the actual methods used are more complicated, we can get a good approximation by doing the following. Divide the current price of each item by its 1967 price and multiply this ratio by the relative weight of the item. This gives us the part of the total CPI contributed by that item. We can do this for each item and add up the products. The result is a ratio of current prices to 1967 prices. To change that ratio to a percent, multiply by 100. This final figure is the CPI.

What does this figure tell us? Since the CPI is technically a percent, if the CPI is currently 250 it tells us that consumer prices are 250% of (two-

and-a-half times as large as) what they were in 1967. But this is not what we usually want to know. We want to know how much prices have increased in the last month, or the last year, or between two other years. For these purposes, we just use the CPI numbers as if they were prices. If bread cost 75¢ a year ago and costs $1.00 today, we know that its price rose by 25¢, or 33%. If the CPI was 200 a year ago and it's 250 today, we know prices rose by 50 points or 25%. People are often confused by the CPI because it is not expressed in terms of dollars. But it is nonetheless a number that stands for the average consumer price at any one time; in this sense, for the year 1967, the average price was 100. This is described by calling the CPI an index in which "1967 = 100."

The Bureau of Labor Statistics releases new CPI figures each month. It releases a bewildering variety of numbers: the CPI itself, the increase over one month ago, over six months ago, over a year ago. One of the figures most often reported by the media is what is known as the annualized rate, which tells what a monthly percentage rate of increase means in annual terms. For example a monthly rise of 1% in the CPI is often expressed as a rise of 12% at an (uncompounded) annual rate.

Since the construction of an index such as the CPI involves a fairly complicated process, even when simplified as we have done here, we will illustrate this process by looking at an actual example of how such an index could be constructed. In the following example, we will use a very small sample of both items and consumers. We will also use simplified techniques for our calculations. However, the basic ideas will be illustrated.

Example 3: Most people budget a certain amount of spending money for lunches, transportation, etc. We will construct a mini price index for spending money of this kind. Using the following data, we will trace through the eight steps outlined above.

1967 Prices

	Jane	Jerry	Joanne	James
Lunch	$ .75	$1.25	$ .25	$ .10
Transportation	0	1.75	.60	1.50
Cigarettes	.60	.45	.15	0
Newspapers	.05	.20	.10	.25
	$1.40	$3.65	$1.10	$1.85

1977 Prices

	Jane	Jerry	Joanne	James
Lunch	$1.25	$2.25	$ .50	$ .30
Transportation	0	3.20	.90	2.75
Cigarettes	1.30	.70	.30	0
Newspapers	.15	.45	.20	.50
	$2.70	$6.60	$1.90	$3.55

1. The base year is 1967.
2. Our sample consists of four office workers: Jane, Jerry, Joanne and James.
3. Our base period market basket includes lunches, transportation, cigarettes, and newspapers.
4. A simplified way of calculating weights is to divide the total for each item by the total for all items.
   a. Lunches: $.75 + $1.25 + $.25 + $.10 = $2.35
      All items = $1.40 + $3.65 + $1.10 +
                  $1.85 = $8.00
      $2.35 ÷ $8.00 = .2975
      .2975 is approximately .30 = weight for lunch
   Notice that the four weights add up to 100%.
   b. Transportation: .48
   c. Cigarettes: .15
   d. Newspapers: .07

Exercise: Do the calculations for b, c and d to make sure you understand the process.

5. We can find the price of the 1967 market basket by simply averaging the four sample baskets.

$$\frac{\$1.40 + \$3.65 + \$1.10 + \$1.85}{4} = \frac{\$8.00}{4} = \$2.00$$

6. Refer to the table above for 1977 prices.
7. We can calculate the ratios for each item, multiply by their respective weights, add, and multiply by 100.
   a. Lunches: 1967: $.75 + $1.25 + $.25 + $.10 = $2.35
              1977: $1.25 + $2.25 + $.50 = $.30 = $4.30
              $4.30 ÷ $2.35 = 1.83 (Lunches cost 1.83 times as
                                    much in 1977 as in 1967)
              (1.83) (.30) =   .549
   b. Transportation: .854
   c. Cigarettes: .288
   d. Newspapers: .152
                             1.843   (1.843) (100) = 184.3

8.   184.3 − 100 = 84.3. The prices in the areas covered by the index have risen by 84.3% between 1967 and 1977.

Exercise: Do the calculations for b, c and d to make sure you understand the process.

The mini price index for 1977 is 184.3. That means $1.84 will buy the same amount of lunches, transportation, cigarettes and newspapers in 1977 that $1 bought in 1967.

The Unemployment Rate. The last economic indicator that we will look at is the unemployment rate. The unemployment rate is the ratio of the number of unemployed persons to the total civilian labor force. The government defines as unemployed anybody between the ages of 16 and 65 who has not worked at a paying job at all during the past week *and* has looked for paid work at some time during the previous four weeks. The labor force is defined as everybody (not in an institution) between those ages who is either working at a paying job (whether full-time or part-time) or who fits the above definition of unemployed. There are problems with these definitions, which we will take up in Part B.

Like the CPI, the unemployment figures are released every month. Also like the CPI, it is the overall figure that is widely reported, although data for a number of subgroups are collected and released. For example, the unemployment figures are broken down into rates for men and women, whites and non-whites, black and white youths (18–25), and heads of households, to name a few. Looking at the figures for these subgroups gives a much better picture of the unemployment situation than does the overall rate.

As it does in computing the CPI, the government uses a sample to obtain data on unemployment. 50,000 households are contacted (approximately 1 of every 1500 households in the country). Every month a quarter of these households are dropped from the sample so that no family is interviewed more than four months in a row.

Questions are asked to determine whether each person of working age in the household worked at all during the last week. If somebody didn't work at all, additional questions are asked to determine whether that person has been recently laid off or is looking for work. If so, he is considered unemployed; if not, he is considered no longer in the labor force, and thus not unemployed.

Example 4: The Bureau of Labor Statistics "Employment Situation" press release for January, 1980 gave the unemployment rate as 6.2%. The total civilian labor force was given as 104,229,000; the number of unemployed persons was given as 6,425,000. Here's how they got the rate:

$$\frac{\text{unemployed}}{\text{labor force}} = \frac{6,425,000}{104,229,000} = 6.2\%$$

Who Is Unemployed?

In general, blacks suffer from much higher levels of unemployment than whites do. For example, in December 1979, the overall unemployment rate was 5.9%. For whites, it was 5.1%; for blacks, it was 11.3%. The difference is even more striking in the youth unemployment figures. Using December, 1979 figures again, for white youths, the rate was 13.9%; for black youths, 34.3%. The unemployment rate for black youths is consistently twice that for white youths. Also, since blacks are disproportionately employed in marginal jobs, they are hit very hard by downturns in the business cycle. Blacks and other minorities bear the brunt of "last hired, first fired" employment policies. This shows up far more clearly when the unemployment statistics are broken down.

Exercise: The figures for 1979, looked like this:
   Total civilian labor force   103,999,000
   Unemployed   6,087,000
   What was the unemployment rate?

Now that we have some idea of how the four main economic indicators, the GNP, the Dow-Jones Average, the CPI, and the unemployment rate are constructed and what they are intended to mean, let's take a closer look at how good a job they do.

PART B: Inadequate Indicators

GNP and Economic Welfare

Many people regard GNP as a measure of people's economic welfare. This can be seriously misleading. The fundamental problem here is that GNP, by measuring only the total value of production, excludes a number of important aspects of people's welfare. We will list a few of the more important issues neglected by GNP.

1. In the first place, by focusing only on products, GNP says nothing about the kinds of jobs that people have or the working conditions on those jobs. There is good reason to believe that one byproduct of business's successful efforts to increase production has been a steady erosion of the quality of working life in American industry.

2. GNP tells us nothing about the condition of the environment. In fact, profit-oriented economic growth has involved more and more air, water, and noise pollution as well as continuing destruction of the countryside.

3. GNP has nothing to say about who gets the goods and services produced. Marketed products are simply valued at their prices. Government services, such as public schools or fire protection, don't have market prices

and are valued instead at what it costs to provide them. Thus, a $100 evening for four at a fancy restaurant enters into GNP as equal to the $100 of groceries that feeds a family of four for two weeks; a $30 hair-styling counts the same as ten $3 haircuts; a $9 million fighter-bomber counts as much as 90 $100,000-per-year child-care centers. When GNP includes such equivalencies, it is hard to regard it as a meaningful indicator of how well off people are. The same level of GNP could provide for very different levels of welfare, depending on how equitably it was distributed among people.

Although the GNP is in itself a fairly simple clearcut measure, it is surrounded with innumerable problems as soon as we try to interpret what it really tells us about the economic state of the nation. When used as a guide to formulating policy, it is not only fraught with difficulties, but with distinct dangers.

Tracking the Stock Market

There are two main kinds of criticisms that can be made of the Dow-Jones Industrial Average. The first has to do with the mechanics by which it is calculated, the second with the ways in which it is interpreted. Since the mysteries of the stock market and its role in the economy of the country are beyond the scope of this book, we will content ourselves with a quick look at the mechanics.

There are three main criticisms that are made of the Dow-Jones Average.

1. The choice of stocks. It is often claimed that the Dow-Jones list contains too many blue chip companies. Blue chip stocks are those which over a long period of time show steady growth and reliable dividends. Excluded are both stocks with less successful records and the glamour stocks which show phenomenal growth due to the successful introduction of a new process or product.

2. The method of calculating the average. Choosing to reduce the number being divided by in order to compensate for stock splits is questionable. When a stock splits, its price goes down, but in a successful stock it will soon drift up again. At any rate, it is important to know that the Dow-Jones Average, as it is now calculated, does not reflect an actual average of the selling price of the thirty stocks on its list. Of course that average in itself is of relatively little interest; it is the fluctuations we care about. The question is the extent to which the changes in the divisor have made the pattern of fluctuations more comprehensible.

3. The use of an unweighted average. As we saw when we calculated the CPI, if we have averaged items of different importance, it is important to weight the averages. In the case of the Dow Jones Average, the stocks

can vary in importance either in terms of volume (the average number of shares sold per day) or in cost (price per share). Neither factor is taken into account. We can see the effect by looking at the following example:

Example 5: On July 29, 1977, two stocks selling at comparable prices were Bethlehem Steel, closing at $22\frac{5}{8}$ and Goodyear, closing at $20\frac{1}{2}$. Bethlehem Steel fell $1\frac{1}{4}$ points, affecting the Dow-Jones by .79 points. Goodyear fell only $\frac{1}{8}$ point, lowering the Dow-Jones average by only .079 points. So far so good. But Bethlehem Steel sold 863,700 shares while Goodyear only sold 32,500 shares, and nowhere is this difference taken into account.

Given its shortcomings, it is surprising that the Dow-Jones continues to be used, not only to keep track of the stock market, but in efforts to predict it. An article explaining the Dow Jones Average in *U.S. News & World Report* of Jan. 19, 1976 said:

"Many stock analysts — known as Dow Jones theorists — use the Dow Jones averages to identify and forecast broad movements in the market. How do Dow theorists do that? Essentially, the concept is simple, but its application is complex and is subject to varying interpretation. It is based on the idea that the market, despite short-term ups and downs, moves in broad trends, lasting sometimes for several years."

Perhaps when you get as general as that, and when you are trying to predict something that is basically unpredictable anyway, the shortcomings of your yardstick don't make that much difference.

Finally, there are other stock market indicators that reflect stock market conditions much better than the Dow Jones Average. Two considered quite superior as indicators are the New York Stock Exchange Index and Standards and Poor's 500.

Keeping Tabs on Inflation

Measuring the true impact of changes in prices on each of the more than 200 million inhabitants of the US is, as we have already said, an impossible task. In this section we will review some of the ways in which the CPI, even though it may be an appropriate measure under certain circumstances, is still inadequate.

Sampling Problems. The first problem we will be looking at is a sampling problem. In Example 3, we used four urban office workers to represent our population. The Bureau of Labor Statistics uses a somewhat larger sample. In the 1960–61 Survey of Consumer Expenditures, 4,343 families were interviewed for the old CPI-W (for wage earners). About 40,000 families were interviewed for the 1972–73 survey when the CPI was revised. One reason that the Bureau of Labor Statistics revised the index is that the CPI-W represented only 40% of the population (urban wage earners and clerical workers). The buying habits of many consumers had been left out. It is obvious that some of these consumers (professionals, retired people) have market baskets that look very different from those of wage earners and clerical workers. If the CPI were simply used as a piece of information to allow people to understand what is happening to their incomes, the individuals sampled would not necessarily be very important. However, the CPI is used to adjust the incomes of many diverse groups. According to a pamphlet put out by the Bureau of Labor Statistics in 1978,

> Collective bargaining contracts covering more than 8.5 million workers tie wages to the CPI. Pensions and other incomes are adjusted to changes in the index for about 50 million social security beneficiaries, retired military and Federal Civil Service employees and survivors, and food stamp recipients. . . . Altogether, a portion of the income of about half the population is pegged to the Consumer Price Index.

Because the CPI is so important in determining increases in wages, the Bureau of Labor Statistics spent 8 years revising the existing CPI as well as generating a new CPI. Despite these changes some sampling problems still remain.

We can illustrate one sampling problem by looking at the fairly homogeneous group of workers in Example 3. As we noted, price changes in basic goods and services affect them differently. Joanne, who uses public transportation, is much less affected by the increase in gas and parking costs than Jerry or James who drive to work; Jane, who walks, is not directly affected at all. The CPI-U took some of these differences into account and changed some of the weights as we mentioned earlier. However, escalating medical costs still affect the elderly more than the young, increases in mortgage interest rates affect buyers of new homes more than long time home owners, and other examples can be cited to show how the CPI still does not provide enough information on price changes for particular groups in the society.

Different parts of the country are also differentially affected. In an attempt to take geographic location into account, the Bureau of Labor Statistics now reports separate CPI's for 28 cities and indexes for four regions (Northeast, North Central, South, and West) by different population groupings. Separate geographical CPI's are useful, but there is no way one index, no matter how carefully constructed, can take into account the different spending patterns of different groups of people who have noticeably different incomes and different life styles.

Weighting Problems. The sampling problem mentioned above involves the use of weights. Some groups will weight items differently than other groups. The kind of weighting problem we are now considering, however, affects all groups. Remember that the present weights for the 400 item market basket were based on a survey taken in 1972–73. Even since 1972–73 buying patterns have changed. This is particularly important in a period of rising prices as in 1978 and 1979. For example, there has for some time been a trend for people to eat more and more of their meals away from home. But, as restaurant prices go up (faster than supermarket prices), people may be shifting back to eating at home more often. In addition, higher gasoline and oil prices may force some groups to alter their buying habits. These shifts in spending preference tend to make the current use of 1972–73 weights questionable.

Weighting, then, to be accurate, would have to be adjusted at much more frequent intervals than it presently is. The research task is so enormous, however, that the amount of time it takes to collect and tabulate the necessary information almost guarantees that it will be out of date before it can be used.

Item Problems. It is not only the weight of items that changes but the items themselves. Over a period of time, people's tastes change, some items cease to be manufactured, and new items come on the market.

Even where the list of items doesn't change, the quality often does. On the one hand, how can you compare the price of a well built older home with some of the less durable current models? On the other hand the quality of television sets has probably improved a great deal in the last 20 years. In both cases, quality changes are difficult to measure.

The CPI, then, is only a very rough indication of how prices are increasing. It is a shoe that probably fits no one but that we are all forced to wear. There is not much we can do about the ways in which it affects our actual income, either directly, as it does for half the population, or indirectly, as it does for all of us when it is used as a guideline to formulate economic policy. (The GNP deflator, used as a measure of inflation by policy makers for some purposes, avoids some of the problems associated with the CPI, but not all of them.) To the extent that we use the CPI as information for our own financial planning, we can make it work better for us by being aware of its shortcomings and also by being alert to how our individual patterns vary from those being charted.

Unemployment: The Rate and the Reality

Because of the definitions it uses in determining the unemployment figures, the government consistently underestimates the true extent of unemployment in the economy. For one thing, the official figures don't include what are called "discouraged workers" — those people who have given up hope of finding a job, and have therefore stopped actively looking. Some studies have indicated that including discouraged workers in the unemployment statistics could raise the rate by more than half. Another problem is that as far as the government is concerned, "part-time is full-time." A person who is working a part-time job, even involuntarily because he can't find a full-time job or because his employer has cut back on hours, is counted as fully employed. This results in another distortion of the real situation.

Example 6:   *Dollars & Sense* magazine, in its November 1974 issue, took the government's official unemployment figure and included estimates for discouraged workers and involuntary part-time workers. Here's what they came up with:

The government's unemployment rate for September, 1974 was   5.8%

Using the government's own very conservative data on the number of people not counted because they have given up looking for work would increase the figure to about   6.8%

But a more realistic estimate of the number of workers not counted because they have given up looking for work would raise the figure still further, to about   8.8%

And a correction for involuntary part-time workers whom the government counts as though they were fully employed would leave us with an unemployment rate for September of   11.8%

There are two other problems that are not measured in the unemployment figures. One is subemployment, the common case where someone with specialized training (say a carpenter or a teacher) is unable to find such a job and has to settle for a less-skilled, lower-paying job.

The other problem with the unemployment figures that we will touch on here is the fact that they don't measure how long a person has been out of work. In recent years, the number of people with at least one week of unemployment at some time during the year has been about four times the number unemployed during an average week. The number out of work for at least a month has been about twice the weekly average. And the number without work for six months or more during the year has ranged between one quarter and one half of the average weekly figure.

The unemployment rate alone, as has been shown, is not an adequate indicator of what's going on in the labor market. And by drastically underestimating the true extent of unemployment, it contributes to political debate based on erroneous information.

SUMMARY

PART A

An **economic indicator** is a number which is used to give information about some aspect of the economy. Four important economic indicators are the Gross National Product, the Dow-Jones Industrial Average, the Consumer Price Index, and the unemployment rate.

The **Gross National Product** is a measure of a nation's output of goods and services. It is the sum of the value of all personal consumption, private investment, net export of goods and services, and governmental purchases of goods and services in one country for one year. It is measured in **current dollars,** but can be adjusted for inflation by using the **GNP deflator** (the ratio of the GNP in nominal to constant dollars). **Constant dollars** measure the value of the same goods and services in terms of what they would have cost in a given year (called the **base period**).

The **Dow-Jones Industrial Average** is a measure of stock prices. It is not really an average, but is the sum of the selling prices of 30 major stocks, divided by 1.465. It is figured continuously and its level indicates the health of the stock market.

The **Consumer Price Index** is a weighted average which measures changes in consumer prices. There are two CPI's: a CPI-U for urban consumers and a CPI-W for urban wage earners and clerical workers. At present the CPI-U and W compare current prices to those of 1967. The latest CPI figures tells us what percent of 1967 dollars we must spend now to buy the same items.

The **unemployment rate** measures the number of unemployed persons as a percent of the total civilian labor force. Released monthly, it gives a general overview of the employment situation.

PART B

Criticisms of GNP include that it measures quantity and not quality, focuses on products not working conditions, and provides no indication as to who gets the goods and services produced.

The Dow-Jones does not provide an accurate picture of stock market behavior because the 30 stocks selected may not be representative stocks, the averaging method is suspect, and stocks selling for different prices and in different volumes have the same effect on the average, per point of change.

The CPI is an inadequate measure of the actual cost of living because it cannot take into account the different spending patterns of different groups of people, changes in buying patterns over time, and changes in the quality of products.

Some of the problems associated with the unemployment rate are that it underestimates the true amount of unemployment by not including discouraged workers and involuntary part-time workers, that it does not measure subemployment, and that it does not provide information about how long people are out of work.

EXERCISES

PART A

1. *If the GNP for 1974 was $1,413.2 billion in current dollars and $1,214.0 billion in constant (1972) dollars, what is the GNP deflator for 1974?

2. If the GNP for 1950 in current dollars was 286.2 billion and the GNP deflator was 53.64, what was the GNP for 1950 in constant (1972) dollars?

3. *If the GNP in 1972 was $1,171.1 billion and in 1973 was $1,306.6 billion with a GNP deflator of 1.058, what was the actual percentage change from 1972 to 1973?

4. The chart below was printed in the December 1974 issue of *Dollars & Sense*. It shows the weights used by the Bureau of Labor Statistics for six main categories of the CPI as well as the rate of inflation for each of those categories in 1974. Try to estimate what portion of your income is spent on each of these categories and complete the chart. By what percent was your dollar devalued in 1974?

CATEGORY	HOW THE BLS DOES IT Aug. 73 – Aug. 74 Weight × Price Increase = _____	HOW YOU DO IT Aug. 73 – Aug. 74 Weight × Price Increase = _____
Food	.248 × 9.0% = 2.23%	_____ × 9.0% = _____
Housing	.333 × 13.0% = 4.33%	_____ × 13.0% = _____
Apparel	.099 × 9.2% = 0.91%	_____ × 9.2% = _____
Transportation	.126 × 15.2% = 1.92%	_____ × 15.2% = _____
Medical Care	.067 × 11.7% = 0.78%	_____ × 11.7% = _____
Recreation & Other	.127 × 8.0% = 1.02%	_____ × 8.0% = _____
TOTAL	1.000 11.19%	1.0 _____

5. *Find the actual percent of price increase for each item in Example 3.

6. Use the method of Exercise #5 to figure the rate of inflation for each of the four people in Example 3.

7. *Which of the people listed below are considered employed? Unemployed? Out of the labor force?
   a. housewife
   b. construction worker who can find only three days' work a week
   c. cab driver with a Ph.D.
   d. student who works part-time
   e. laid-off auto worker
   f. full-time factory worker

PART B

1. Suppose you wanted to construct an index of economic growth that took quality into consideration as well as quantity. What categories would you use? Would you make the index a sum, an average, or a weighted average? Why?

2. Use either the high and low values for the Dow-Jones found in Exercise 4 above, or use some more recent highs and lows. Read one or two news magazines for the period and see if you can explain the activity of the stock market in terms of concurrent political or economic developments.

3. *Look again at Exercise 4 above. Name some of the ways in which the rate of inflation corresponding to your weights still did not accurately measure your personal cost of living increase for the year 1974.

4. Look back at Part A, Exercise 7. Which of these people fit into the categories we developed in Part B?

Chapter 5

FREQUENCY DISTRIBUTIONS

PART A: Grouping Data

In statistics we often have to deal with very large amounts of data. In order to make sense out of hundreds of test scores, or thousands of responses to an opinion poll, we have to find some way to group the data. Instead of looking at a thousand individual responses to the question, "What kind of a job do you think the President is doing?" we may want to group those responses together under three headings — Good, Fair, and Poor — or maybe under five headings — Excellent, Good, Fair, Poor, and No opinion. Then we will want to know how many responses there are in each group. We call each of the groups we use a **class;** the number of responses which fall into each class is called the **frequency** of that class; and when the data has been grouped into classes and the frequency of each class has been counted, the result is called a **frequency distribution.**

Frequency Distributions

Example 1: Suppose we give a test to 50 students. Then we score each test on a scale of 1 to 10. The test scores are:

2, 10, 6, 5, 9, 4, 7, 8, 5, 8, 9, 7, 10, 6, 8, 5, 7,
9, 1, 10, 9, 10, 5, 2, 2, 4, 6, 8, 7, 7, 9, 5, 6, 1,
5, 7, 9, 8, 9, 10, 5, 10, 7, 4, 7, 8, 4, 9, 8, 9

We want to make a frequency distribution of these scores so that we can get a better idea of how our students did. To do this, we first set up three columns. Then, in the first column, we identify the classes that will be used. In the second, we count, by tally, the number of scores in each class. In the third, we convert each tally to a number which represents the frequency for that class. Finally, we add up the numbers in the last column to make sure all the scores were counted. The total should be the number of scores we began with — in this case 50. The frequency distribution for these 50 scores will look like this:

Frequency Distribution of Scores, #1

Score	Tally	Frequency
1	11	2
2	111	3
3		0
4	1111	4
5	⊬⊬⊦ 11	7
6	1111	4
7	⊬⊬⊦ 111	8
8	⊬⊬⊦ 11	7
9	⊬⊬⊦ 1111	9
10	1111 1	6
		50

Frequency Histograms

The above table tells us how many students scored 1, how many scored 2, and so on up to 10. It tells us how the scores are distributed among the 50 students. A better way to get a picture of the distribution, however, is to make a graph. Any of the graphs discussed in Chapter 3 could be used, but in practice frequency distributions are usually pictured by either line or bar graphs. If we use the bar graph, we construct a **frequency histogram.** This is a bar graph in which the bars are vertical and are usually placed together so that there are no spaces between them. The horizontal scale records the scores, or classes of scores, and the vertical scale the frequencies. Figure 1 is a frequency histogram of the scores in Example 1.

If you have difficulty understanding the histogram, try visualizing a situation like this: Suppose we place cards numbered from 1 to 10 along a table. We then ask each of the 50 students from Example 1 to line up behind the number which represents his or her score. Now imagine hanging from the chandelier and looking down on all those heads. What you will see can be thought of as a life-sized histogram. This example is shown in Figure 2.

Example 2: Suppose now that we give the same 50 students another test, but that this time we score it on a scale of 0 to 100. The scores are:

98, 87, 57, 84, 66, 69, 75, 78, 86, 95, 68, 92, 97, 64, 100, 84, 56, 85, 94, 78, 72, 61, 96, 65, 70, 74, 80, 90, 93, 83, 87, 59, 73, 62, 94, 84, 77, 64, 79, 92, 87, 79, 62, 75, 71, 90, 94, 76, 72, 61

Figure 1

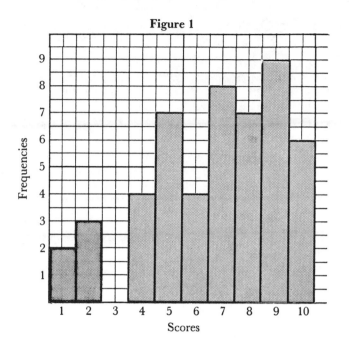

Figure 2

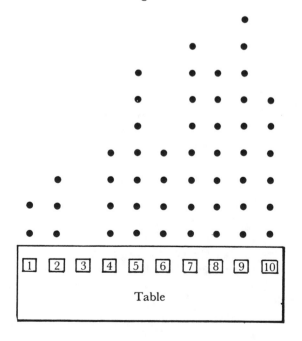

Table

In Example 1, each score formed a class of its own. It would make little sense to do that for the scores in Example 2. Since the scores range from 56 to 100, we would need almost as many classes as scores. The whole point of grouping together many scores in a few classes would be defeated. So we must decide how many classes to use and what they will be. There is no rule about this; we just want a reasonable number of classes. If there are too few scores per class, we may as well use the raw data. If there are too few classes, we will lose the significance of the different scores. Common sense and experience are the best guides we have. But no matter how many classes we use, there are two rules we must keep in mind when we set them up.

1. There must be the same number of possible scores in each class. That is, if we think of each class as a category, these categories must all be the same size, even though the number of actual scores in each category (the frequencies) will of course be different from one class to the next.

2. There must be no overlap between classes. The **boundary points,** (the first and last number in each class) must be distinct. If the first class contains scores 1–10, the second class must start with 11, so we will have 1–10, 11–20, 21–30, etc. Or alternatively, 1 to under 10, 10 to under 20, 20 to under 30, etc. We *cannot* have categories like 1–10, 10–20, etc. because then there would be two possible classes to contain the scores 10, 20, etc. Each score must fit into one and only one class.

So let's see how we find the class boundaries once we have decided how many classes we want. Suppose we decide to use 8 classes. What will their boundaries be? First we need to know how many possible different scores there are. Since the lowest score is 56 and the highest 100, we find that: $100 - 55 = 45$. There are 45 possible different scores. Notice that we subtract 55 instead of 56 from 100 since 56 is included in the group of possible scores. Now to find the size of each class, we divide 45 by the number of desired classes.

$$45 \div 8 = 5.625$$

We want to avoid fractions, which will make awkward class sizes, so we have two choices; we can use either 5 or 6 as our class size. If we use a class size of 5, we will need 9 classes ($5 \times 9 = 45$). If we use a class size of 6, we will have 8 classes, but we will cover a larger range than we need ($6 \times 8 = 48$). If we decide to use 8 classes with 6 possible scores in each class, we will simply have to start our first class with a smaller number (53) than our lowest score (56). On the whole it seems more convenient to use 9 classes.

Frequency Distribution of Scores, #2

Class	Tally	Frequency
56–60	111	3
61–65	++++ 11	7
66–70	1111	4
71–75	++++ 111	8
76–80	++++ 1	6
81–85	++++	5
86–90	++++ 1	6
91–95	++++ 11	7
96–100	1111	4
		50

There are two ways we can indicate the classes along the horizontal axis. One way is to label the boundaries between classes, as in Figure 3. Alternatively, we can label the midpoint of each class, as in Figure 4. Notice that the **midpoint** of a class is the point halfway between the left boundary point and the right boundary point.

Figure 3
Frequency Histogram:
Boundaries Labeled

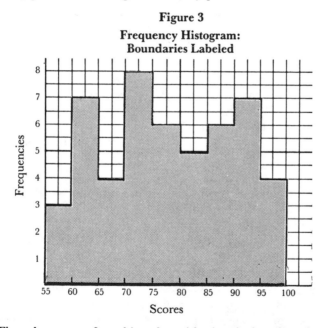

The advantage of marking the midpoints is that then there is no confusion about where scores that fall on the boundary points are recorded, in the class before or after the point. However, as long as we are consistent, this need not be too much of a problem. From the frequency distribution it is clear that in the histogram above, the score corresponding to each boundary point is included in the class to the left of the number, not the class to the right.

Figure 4

Frequency Histogram:
Midpoints Labeled

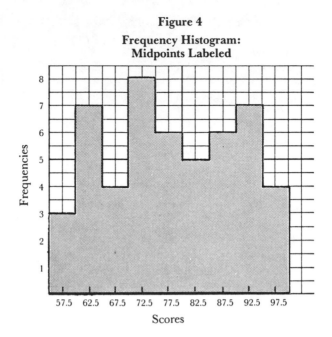

Scores

Exercise: Construct a frequency distribution and frequency histogram based on the gross weekly wages of secretaries at ABC Insurance company in 1979.

110, 135, 115, 185, 125, 85, 115, 140, 110, 95, 100, 175, 150, 105, 120, 90, 90, 130, 110, 165, 80, 140, 110, 110, 120, 190, 100, 95, 130, 135

Relative Frequency Histograms

There are also two ways of marking the vertical axis of a histogram. We can either indicate the absolute frequencies, as we have done so far, or we can indicate the relative frequencies. The **relative frequency** of a score is simply the absolute frequency divided by the total number of scores. If there are 50 scores and 3 of them are in the class 56–60 then that class has a relative frequency of $\frac{3}{50}$; that is, the class contains $\frac{3}{50}$ (three-fiftieths) of all the scores. The relative frequency can also be expressed as a percent, so we could say that 6% of the scores fall in the class 56–60. Notice that the relative frequency histogram in Figure 5 looks exactly like the frequency histogram except for the numbers along the vertical axis. Instead of every two squares along that axis representing one test score, as they do in Figure 4, here that same distance represents $\frac{1}{50}$ (or 2%) of the total number of scores.

Figure 5

Relative Frequency Histogram

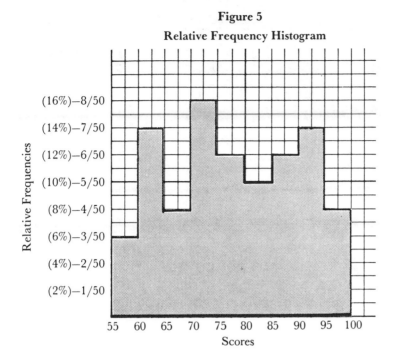

Relative frequencies are much more important than absolute frequencies, especially when we are dealing with a large number of cases. If we look at the test scores for say all 5th graders in the city of Boston, we are much more interested in knowing what percent of the group placed above grade level than in knowing the actual number of students that did. Relative frequency histograms help us to see this at a glance, because the relative frequency of a class is exactly equal to the relative area occupied by that class compared to the total area of the graph. By "total area of the graph," we mean the total area taken up by the bars, not the total amount of space inside the axes. All histograms have the property that the area of the graph occupied by a given class has the same ratio to the total area of the graph as the number of scores in that class have to the total number of scores. This ratio is the relative frequency.

Figure 6 shows how the area of the graph for the data in Example 2 can be divided up into 50 squares. The area of the first class is made up of 3 of those squares, the second class of 7, the third class of 4, and so on. And the relative frequencies of those classes are $3/50$, $7/50$, $4/50$, etc.

Figure 6

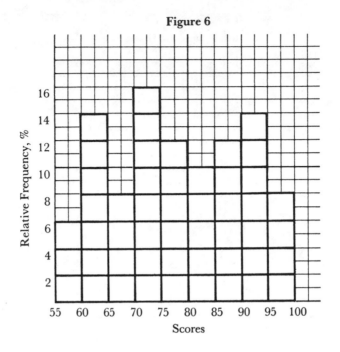

Frequency Polygons

As we said earlier, we can use either a bar graph (histogram) or a line graph to display frequencies. If we use the line graph, we have a **frequency polygon.** Instead of making bars, we place a dot over the midpoint of each class to represent the frequency of that class. Then we connect the dots with a line. The line is extended to the base line, or horizontal axis, at each end of the graph. Figure 7a shows the frequency polygon for the data in Example 2. Polygons, like histograms, can measure either frequency or relative frequency, and again the only difference is in the labeling of the vertical scales. Compare Figure 7a, the frequency polygon, with Figure 7b, the relative frequency polygon.

Remember that in a histogram the enclosed area of each class is proportional to the relative frequency of that class. This is also true of a polygon. Look at Figure 8. The shaded area encloses the area of the class of scores falling between 80 and 85. We see from the frequency distribution of p. 101 that this class contains 5 of the 50 scores or 10%. Now let's see what the corresponding area on the graph is.

Figure 7

a. Frequency Polygon

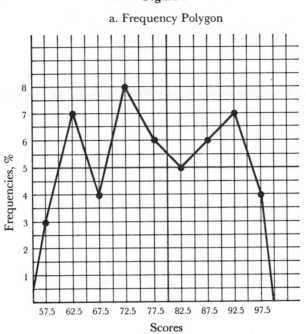

b. Relative Frequency Polygon

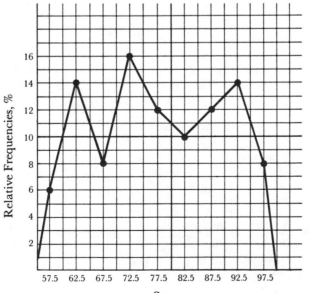

Figure 8

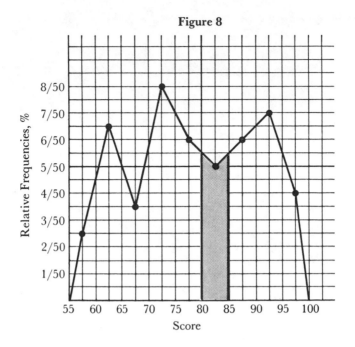

We begin by marking off the boundaries between classes rather than the midpoints of each class on the horizontal scale. If we then count the total number of squares on the graph paper enclosed by the line of the graph and the horizontal scale, we find that there are approximately 200 such squares. Next we count the number of squares occupied by the shaded area, the class of scores between 80 and 85. We find that there are approximately 21 squares in that area. So the ratio of that area to the total area is $^{21}/_{200}$ or approximately 10% — the relative frequency of the scores in that class. The importance of this relationship, between relative frequencies and ratios of areas, cannot be stressed too strongly.

Exercise: Estimate the number of squares occupied by the area which is shaded in Fig. 9 below. This area is made up of three classes. What percent of the total area is occupied by these three classes? What is the sum of the relative frequencies of these three classes? Are these numbers approximately equal?

Up until now, all our examples of frequency distributions have used test scores. But frequency distributions, histograms, and polygons can be used to represent many types of data. The wages received by different workers, as in our exercise with the secretarial wages, would be one example. Others include different ages of the workforce in a factory, different heights of school children, etc. In addition, we have used the word "scores"

Figure 9

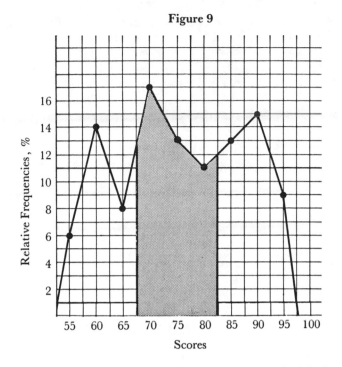

to refer to the data whose frequency we're concerned with, because these data have in fact been test scores. In future chapters, however, we'll often use the word scores to represent all "data points" such as inches (if we're concerned with the frequency with which different heights occur), dollars, etc.

PART B: Questionable Groupings

Like any other statistical technique, grouping data into frequency distributions can cover a multitude of sins. Depending on the size of the classes or the number of classes and their beginning and end points, the appearance of a distribution can be drastically changed.

Here is a commonly used example:

Example 3: A quiz is administered to 35 students and graded on a scale from 1 to 10. The results are as follows:

Score	10	9	8	7	6	5	4
Frequency	3	5	7	8	5	4	3

Suppose now that these scores are to be changed to letter grades. How many classes do we need? If the possible grades are A, B, C, D and F, we need 5. If we decide that everyone passed, we can get by with 4, and if the quiz was sufficiently difficult, we might decide the lowest grade should be a C, in which case we only need 3 classes. Should the classes all be the same size? Probably not, for the purpose of assigning grades, but suppose we do decide to do it on the basis of a strict frequency distribution. If we decide the lowest grade should be a C, we will have to find three classes for the 7 different scores. But $7 \div 3 = 2\frac{1}{3}$. We can't use a fraction because we can't break up one of the scores into two classes. That is, we can't give one person who scored 9 an A and another with the same score a B. Therefore the class size has to be rounded off to 3. Now since we have 3 classes of 3 to distribute 7 possible scores into, we have a good deal of leeway as to how we do it. Here are the possibilities:

Figure 10

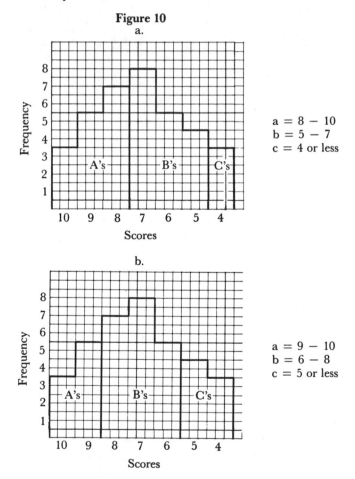

a.

a = 8 − 10
b = 5 − 7
c = 4 or less

b.

a = 9 − 10
b = 6 − 8
c = 5 or less

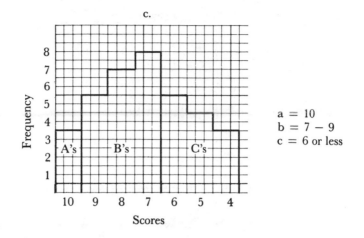

c.

a = 10
b = 7 − 9
c = 6 or less

In the above example only the people who score 10, 7, or 4 got the same grade in all three cases. Also, note that in our possibility "a" we are giving 43% of the students A's and only 9% C's whereas in possibility "c" we are giving 9% A's and 34% C's. In other words, using the technique of frequency distributions, an "objective" statistical technique, a teacher could make grades come out almost any way she pleased.

Although it is sometimes impossible to make our classes the same size, as in the example above, we can get very misleading results when we don't. For example, we can give the impression of an effect that doesn't really exist. Suppose a toothpaste advertiser wants to make her product look good. She knows that the average child using any toothpaste may be expected to acquire two new cavities every six months during the growing years. Children using her toothpaste are no exception. But by selecting classes appropriately, it is possible to give the impression that the average is expected to be much higher and that users of this product are benefiting.

Example 4: The raw data are:

Number of cavities

per 6 month period:	0	1	2	3	4	5	6 or more
Number of children:	5	45	60	37	10	4	3

By grouping the first three categories together into one large class, while making each of the other categories into a single small class, we can get a picture like this:

Figure 11

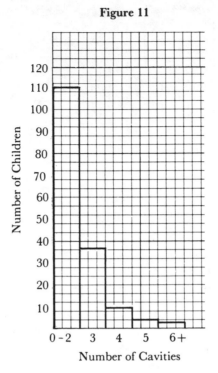

This graph can be accompanied by a statement which says "users of our toothpaste had fewer cavities. Over 70% had only 2 cavities or less."

In example 4, we see grouping being used to give the impression of an effect that does not exist. Inept groupings can also obscure real phenomena. Example 5 is a hypothetical example of this kind.

Example 5: Suppose we wanted to check out the belief that more crimes are committed while the moon is full than at other times of the month. We look at records for one lunar month to see if further investigation is warranted. Day #1 is the first day of the new moon while day 28 is the last day before the next new moon, so the full moon is on the night between day #14 and day #15. Here is what we find:

Days:	1	2	3	4	5	6	7	8	9	10	11	12	13	14
Number of Crimes:	2	3	1	3	0	4	2	2	3	1	0	2	3	6

Days:	15	16	17	18	19	20	21	22	23	24	25	26	27	28
Number of Crimes:	7	5	3	1	0	0	1	3	1	4	3	2	1	2

We have twenty-eight days of data to deal with. The conventional way of dividing a lunar month is into four "phases" of seven days each. Suppose we use these standard phases of the moon for our classes.

New Moon		1st Quarter		Full Moon		Last Quarter	
Day	# of crimes	Day	# of crimes	Day	# of crimes	Day	# of crimes
1	2	8	2	15	7	22	3
2	3	9	3	16	5	23	1
3	1	10	1	17	3	24	4
4	3	11	0	18	1	25	3
5	0	12	2	19	0	26	2
6	4	13	3	20	0	27	1
7	2	14	6	21	1	28	2
	15		17		17		16

The differences between the numbers of crimes committed during each lunar phase are quite small. The "full moon" phase does not correspond with significantly more crimes than any other. We label the lunar theory of crime moonshine and forget it.

But we have made a serious error in our grouping. Perhaps we have been misled by the conventional names of the phases and have failed to notice that the "full moon" phase is really the phase which *follows* the full moon. For whatever reason, we have divided the actual period of the full moon between two large classes. Suppose instead we had decided to group the lunar month into seven classes of four days each. Here is what we would have found:

Day	# of crimes	Day	# of crimes	Day	# of crimes	Day	# of crimes
1	2	5	0	9	3	13	3
2	3	6	4	10	1	14	6
3	1	7	2	11	0	15	7
4	3	8	2	12	2	16	5
	9		8		6		21

Day	# of crimes	Day	# of crimes	Day	# of crimes
17	3	21	1	25	3
18	1	22	3	26	2
19	0	23	1	27	1
20	0	24	4	28	2
	4		9		8

What conclusions would you draw from this grouping of the data?

Frequency distributions are only one kind of grouping of data, and as long as the classes are selected carefully, the results are likely to be fairly clear and free of deception. Other kinds of groupings, particularly those relating to qualitative data, have a variety of dangers attached to them. We will only give one example here, but you will no doubt be able to think of others.

Our final example will look at a situation where grouping is used to create a false impression through non-numerical manipulations.

Example 6:   In the September 11, 1974 issue of *The Real Paper,* Robert F.
Duboff writes about some of the drawbacks of public opin-
ion polls. In this article he mentions the following interesting fact: "Professor Hadley Cantril once followed up a
Gallup survey where three choices had been supplied. Cantril found that only 3 people out of 40 volunteered any of the
three answers offered by pollsters when they were given free
rein to answer as they chose."

The point is that strictly limiting the number of allowed responses to
a question (or forcing a variety of responses into few categories at some
later stages of tabulation) is likely to distort people's real opinions even
under the best of circumstances. If the purpose of the poll is to make a
candidate or product look good, it is not hard to see how ambiguous answers can be classified as desired.

The conclusion to be drawn from the above examples is that although
frequency distributions and other forms of grouping are useful and necessary to allow us to draw conclusion from large amounts of data, they also
present dangers from both inept and unscrupulous users. Although it is
not always possible to get access to the raw data, it is often worth considering what one would expect it to look like, and in very important cases to
make the effort to get hold of it.

SUMMARY

PART A

Large amounts of data must be grouped in order to become meaningful. One type of grouping is called a **frequency distribution.** The data is
grouped into **classes** and the number of items in each is called the **frequency** for that class. Classes must cover the range of possible data points
without gaps or overlapping and must be of equal size. Either the **midpoint** (halfway between the boundary points) or the **boundary point** may
be labelled.

Frequency distributions can be graphed as bar graphs, which are
called **frequency histograms,** or as line graphs, called **frequency polygons.**
The horizontal axis always represents the classes, the vertical axis the frequencies.

The frequencies recorded along the vertical axis can be absolute
frequencies, or **relative frequencies** (the percent or fraction the class frequency is of the total of all the frequencies).

The ratio of the area of a graph occupied by one class to the area
occupied by the total graph is always equal to the relative frequency of
that class.

PART B

Frequency distributions can give different impressions depending on the size and number of the classes. In cases which come out uneven, the choice of beginning and end points for the first and last class can also affect the total picture.

Groupings other than frequency distributions also give opportunities for manipulating data on the basis of which categories are set up and how specific data is assigned to them. The most frequent problems are:

   a. effects that don't exist but are created by false groupings
   b. effects that do exist but are obscured by inept groupings
   c. non-numerical manipulations

EXERCISES

PART A

1. *Use the following test scores to make your own frequency distribution. Use whatever number of classes seem reasonable to you.

   Scores: 66, 98, 75, 97, 78, 87, 89, 65, 74, 82, 86, 76, 78
           73, 79, 91, 95, 89, 81, 63, 80, 96, 88, 90, 100

2. Draw a frequency histogram and a frequency polygon for the above data.

3. *Draw a relative frequency histogram and polygon for the above data, using graph paper.

4. What is the relative frequency of your last two classes? Count graph paper squares to see what the total areas under the graph and under the last two classes are. Is the ratio of these two numbers the same as the relative frequency?

PART B

1. *What letter grades would you assign to the scores in Exercise 1? Why?

2. A poll is taken to see how people of different age groups feel about Candidate Jones. 100 people in each age group are asked if they would vote for him and the results are as follows:

Age Group:	under 20	21–30	31–40	41–50	51–60	over 60
Favorable Responses:	5	25	60	50	35	15

   a. How would the pollster for Candidate Jones regroup this data to make it appear that the candidate was equally popular with young and old alike?

   b. *How would the pollster for Candidate Jones regroup it to show that only the immature were opposed to Jones?

   c. How would Candidate Jones' pollster present the fact that the mature (but not senile) section of the population was squarely behind his candidate?

3. If you could look at the raw data from which the following report was drawn, what questions would you ask? What would you look for?

The following is quoted from *Time*, March 17, 1975.

ENERGY TAXES. Despite the low confidence in political leaders, people clearly want action from them on energy and economic policy. By 55% to 34%, those polled preferred gasoline rationing to a tax that would increase gasoline prices. Even Republicans looked with more favor on rationing than on proposed efforts to cut consumption through higher prices (45% *v.* 43%), while Democrats backed rationing by more than 2 to 1.

To stimulate the economy, 78% of those polled thought limiting a tax rebate to lower- and middle-income taxpayers would be best, while only 19% supported Ford's proposal of an across-the-board rebate up to a maximum of $1,000. There was strong sentiment for tax incentives for business to help the economy; 78% favored the idea, up eight points from last fall. By heavy majorities, people favored bringing back wage-price controls (69% agreed), cutting defense spending (72%), reducing foreign aid (82%) and loosening credit and mortgage money (82%).

Chapter 6

AVERAGES

PART A: Finding Typical Values

In Chapter 4 we saw that averages, either simple or weighted, can be used as single number summaries of large amounts of data. We used a simple average to find the price of milk in a given year and we used a weighted average to calculate something like the Consumer Price Index. We also mentioned in that chapter that statisticians use a variety of calculations that result in what might be called averages, of which these were just two examples. We are now ready to look at averages from a statistical point of view.

The kind of average we have been talking about so far, where we add up the items and then divide by the number of items, with or without the use of weights, is what statisticians call the **arithmetic mean** (simple or weighted) or just the **mean.** The word **average,** if it is used in statistics at all, refers to the result of any one of a number of different calculations, each designed to give us a different view of what the "typical" value of a set of numbers might be. The mean is one of these. In this chapter we will also look at two other commonly used averages, the median and the mode. These and other averages are often called measures of "central tendency."

The Mean. To find the mean, we add up all the numbers in our set and divide by however many there are. Suppose we have the following test scores we wish to average.

Example 1: 1, 2, 2, 2, 5, 5, 6, 6, 6, 6.
$1 + 2 + 2 + 2 + 5 + 5 + 6 + 6 + 6 + 6 = 41$
There are 10 scores. $41 \div 10 = 4.1$
The mean of these scores is 4.1

Beginning with this section, we will be using a variety of formulas to calculate certain characteristics of distributions. A formula is simply a shorthand way to write down what operations we need to perform in order to get a certain result. In the case of the mean, we can say in words: "To find the mean of a set of numbers, add all the numbers together and then divide the sum by the number of numbers in the set." Or we can say the same thing with this formula:

Formula 1: $$\bar{x} = \frac{\Sigma\, x}{n}$$

Formula 1 is much shorter and more exact than the sentence, but it is more difficult to read until the symbols become familiar. Here's what they mean:

$\bar{x}$ (read x-bar) stands for the mean.

Σ (the Greek letter, capital sigma) stands for sum. It does not represent a number, like other algebraic letters do, but rather an operation. Just as a "+" sign tells us to add together the numbers surrounding it, the "Σ" sign tells us to add together what comes after it.

x in this formula stands for not one number, but all the numbers that form the set or distribution whose mean we are calculating.

n stands for the number of items in the set.

It is worthwhile memorizing the meaning of these symbols because we will continue to use them throughout our study of statistics. We will always use x's to designate the data we are working with. (When we have two sets of data we will use y's for the second set.) A letter with a bar over it will always indicate the mean of the numbers represented by that letter. The letter n will always represent the number of data points in the set, and the capital sigma (Σ) will always tell us to add.

Now that we know what the symbols refer to, we need to know the order in which the operations they indicate must be performed. The order is the same as it always is in algebra. A fraction is an indication to divide, but we cannot carry out the division until all operations in the numerator and denominator have been performed. Since Σ in the numerator tells us to add, we must carry out that addition before we divide. So we add up all the x's and then divide by n to get the mean.

Notice that unlike the formulas you are used to in algebra and geometry, you can't use this formula by substituting single numbers for the letters. In geometry, the formula $A = bh$ means that to find the area of a rectangle you multiply the base times the height. If you have a rectangle that is 3' by 5' you substitute 3 for b and 5 for h to get

$A = (3)(5) = 15$. The area of the rectangle is 15 square feet.

But with the formula

$$\bar{x} = \frac{\Sigma\, x}{n}$$

that kind of simple substitution doesn't work. We can't substitute one of the scores for the x. Instead, we have to think of the formula as a short-hand form of directions, not as an equation into which we substitute numbers for some of the letters and solve for the remaining variable.

The formula we have been examining tells us how to find the mean for **ungrouped** data. Now let's see how we must change it to handle data **grouped** into a frequency distribution. First, let's group the data from Example 1:

Scores (x's)	Frequencies (f)
1	1
2	3
5	2
6	4
	10

If we think of multiplication as a shortcut for addition, we can see a quick way to find the mean of grouped data. Instead of adding four 6's, we can multiply (4)(6); instead of adding two 5's, we can multiply (2)(5), and so on. Then we add the results of our multiplications and divide the sum by the total number of scores, just as we did before.

Scores (x's)			Frequencies (f)		Frequencies times Scores (fx)
1	multiply by		1	=	1
2	"	"	3	=	6
5	"	"	2	=	10
6	"	"	4	=	24
			10		41

$$\bar{x} = \frac{\Sigma\, x}{n} = \frac{41}{10} = 4.1$$

The mean is 4.1, the same result we got by adding all the scores and then dividing in Example 1.

Now let's rewrite the formula for finding the mean of grouped data. What did we do first? We multiplied each different score (each x) by its frequency (which we can call f). Algebraically, we can write that multiplication as fx. Now we want to add up all those products. Remember that the symbol for addition is Σ, so we can write our sum as $\Sigma\,(fx)$. The parentheses tell us that we must do the multiplication before the addition. Finally, we again divide by n and our new formula is:

Formula 2: $$\bar{x} = \frac{\Sigma\,(fx)}{n}$$

Be sure you understand what the *n* stands for. It represents the total number of scores, or *x*'s. It represents the sum of the frequencies. It does *not* represent the number of classes. In this example then, n = 10, *not* 4.

The example we used to develop this formula was a distribution in which each class consisted of only one number. What do we use for *x*'s when that is not the case? Look at the following example:

Example 2:   1, 1, 2, 3, 3, 4, 4, 4, 4, 5, 5, 6, 7, 8, 8, 9, 9, 9, 9, 9.

Class	Frequency
1–3	5
4–6	7
7–9	8
	20

Remember that when we were constructing frequency polygons, we used the midpoint of each class to represent the whole class. We can use the same solution here. In the frequency polygon we placed the dot representing each class over the midpoint. Here we use the value of the midpoint (*x*) of each class in place of the *x*'s in our formula.

Class	Class Midpoint (*x*)	Frequency (*f*)
1–3	2	5
4–6	5	7
7–9	8	8
		20

$$\bar{x} = \frac{\Sigma\,(fx)}{n} = \frac{(5)(2) + (7)(5) + (8)(8)}{20} = \frac{10 + 35 + 64}{20} = \frac{109}{20} = 5.45$$

The mean is 5.45.

Notice, however, that when we calculate the mean directly from the ungrouped data, we get a slightly different answer:

$$1+1+2+3+3+4+4+4+4+5+5+6+7+8+8+9+9+$$
$$9+9+9 = 110 \quad 110 \div 20 = 5.5$$

This should not surprise us, since using the midpoint of the class to represent the whole class is an approximation that may sometimes be quite far off the mark. Look at the last class. The midpoint is 8 and we have eight *x*'s in that class, so *fx* = 64. But if we actually look at the class we see that there are five 9's and only two 8's and one 7, so the real sum of the *x*'s is 68. Usually when we use grouped data, the number of items is so large that the discrepancy is small enough not to affect the results, but it is important to know that when using this method to find the mean, the answer must be considered an approximation, not the exact mean.

Finally, we should note that we do not really need two formulas for the mean, one for grouped and one for ungrouped data. The formula for grouped data will give us the correct answer in both cases. If the data is ungrouped, we simply take the value of f as equal to 1 in each case. Example 3 illustrates this point.

Example 3:   1, 9, 6, 5, 7.

Formula 2: $\bar{x} = \dfrac{\Sigma (fx)}{n} = \dfrac{(1)(1) + (1)(9) + (1)(6) + (1)(5) + (1)(7)}{5} =$

$$\dfrac{1 + 9 + 6 + 5 + 7}{5} = \dfrac{28}{5} = 5.6$$

Formula 1: $\bar{x} = \dfrac{\Sigma x}{n} = \dfrac{1 + 9 + 6 + 5 + 7}{5} = \dfrac{28}{5} = 5.6$

Exercise: Find the mean of the data below using Formula 1 and 2.
         4, 8, 10, 2, 5, 1, 5, 11, 3, 6, 2, 9, 5, 7, 10, 3, 12, 6, 2, 7.

The Median. Though the mean is the most familiar average, the median is actually simpler to calculate. It is the mid-point of the whole distribution. To find the median, you line up all the scores in the previous example from smallest to largest (or largest to smallest, it makes no difference), and find the middle one. By "the middle one" we mean the score which has the same number of scores above and below it. If your distribution contains an odd number of scores, there is a natural middle one.

Example 4:   3, 6, 2, 6, 4, 3, 7
         These scores must first be rearranged from largest to smallest or smallest to largest.
              7, 6, 6, 4, 3, 3, 2 or 2, 3, 3, 4, 6, 6, 7
         In either case it is obvious that 4 is the median because there are the same number of scores higher than 4 as lower than 4.
         7, 6, 6, 4, 3, 3, 2   2, 3, 3, 4, 6, 6, 7   The median is 4.

When there is an even number of scores in the distribution, there is no one middle number, so we take an average of the middle two numbers.

Example 5:   2, 3, 3, 5, 6, 6, 7, 7
         This distribution divides evenly into two groups of four
                    2, 3, 3, 5, 6, 6, 7, 7
         In this case we take the two middle numbers (the largest of the low group and the smallest of the high group) and average them.
              $\dfrac{5 + 6}{2} = 5.5$.   The median is 5.5.

Suppose now that we had a distribution of 99 scores. After they are arranged according to size, we can pick out our median as the value of the 50th item. It will have 49 items above it and 49 below.

Exercise: How do we pick out the median from 100 scores?

We pick out the median from grouped data in exactly the same way, except that instead of using actual scores we use midpoints of classes just as we did in finding the mean from grouped data.

Example 6:

Class	Midpoint	Frequency
1–3	2	6
4–6	5	6
7–9	8	8
		20

We can see that since there are 20 scores, the median is the average of the 10th and 11th ones. Both of these scores are in the middle class (4–6), so our median is in that class. Of course we can't use a class interval for a median. Therefore we use the class mid-point and say that our median is 5. If the two middle ranks in an even numbered distribution fall into two different classes, then we just average the midpoints of those two classes.

Example 7:

Class	Frequency
51–60	5
61–70	7
71–80	10
81–90	13
91–100	9
	44

Since there are 44 scores, the median is the average of the midpoint of the classes containing the 22nd and 23rd ones. These are the classes 71–80 and 81–90. The midpoint of the first is 75.5 and of the second 85.5.

$$\frac{75.5 + 85.5}{2} = 80.5. \quad \text{The median is 80.5.}$$

The Mode. The mode is the easiest of all the averages to find. It is simply that number in a distribution which occurs most frequently.

Example 8:   2, 3, 3, 4, 4, 5, 5, 5, 5

5 is the mode of this distribution because there are more 5's than any other number.

In the case of grouped data, the mode is the midpoint of the class with the highest frequency. Look at Example 9.

Example 9:

Class	Frequency
3–5	23
6–8	5
9–11	160

In this example the mode is 10.

Sometimes there are two scores or classes which occur with the same highest frequency.

Example 10:   2, 3, 3, 3, 4, 4, 5, 5, 5

In this example there are three 3's and also three 5's. Both 3 and 5 are considered modes of this distribution.

Distributions with two modes are called **bimodal distributions.** If there are more than two scores with the same highest frequency then the idea of a mode ceases to have much meaning; such distributions are called **multimodal.**

Example 11: 2, 2, 3, 3, 3, 4, 5, 5, 5, 6, 6, 6

This is a multimodal distribution because there are three 3's, three 5's and three 6's.

Exercise: Find the mean, median and mode of the following distribution: 8, 10, 25, 16, 14, 23, 10, 21, 28, 25, 7, 16, 5, 15, 20, 17, 12, 10, 14, 21, 19, 26, 12, 11, 22, 5, 15, 10

The Appropriate Average

Up to this point we've looked at three different kinds of averages, the mean, median and mode, in terms of distributions of abstract scores. Each is supposed to tell us the "typical value" of a distribution. Which is best for representing what is typical in real world situations, like a typical height or income or heart rate? There is no one answer to that question, but we can ask how the three measures differ and then examine their relative advantages and disadvantages in various situations. First, let's examine a case where the three measures all produce the same result.

Example 12: Suppose we measure a group of 100 men to see how tall they are.

Class	Midpoint (x)	Frequency (f)	(fx)
60"–62"	61"	3	183
63"–65"	64"	6	384
66"–68"	67"	24	1608
69"–71"	70"	36	2520
72"–74"	73"	22	1606
75"–77"	76"	8	608
78"–80"	79"	1	79
		100	6988

Mean: $\bar{x} = \dfrac{\Sigma (fx)}{n} = \dfrac{6988}{100} = 69.88$ or 70"

Median: The 50th and 51st measures are both in the class whose midpoint is 70"

Mode: The midpoint of the class with the highest frequency is 70"

In this example, then, the mean, median and mode are all the same: 70″. Whenever the three measures are identical, or very close together, we know not only that it makes no difference which measure we choose, but more importantly, that we have a very special kind of frequency distribution called a symmetrical distribution. It's special characteristics can best be seen by looking at its frequency histogram (Figure 1a) and frequency polygon (Figure 1b):

Figure 1

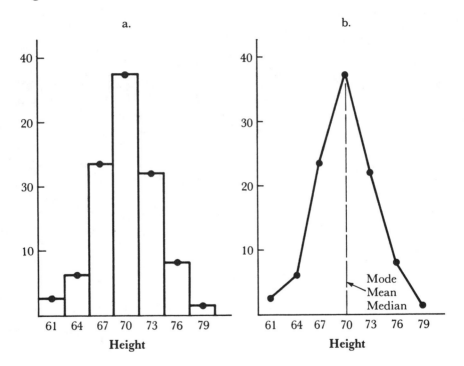

We say a distribution is symmetrical when the two halves are identical. In this case the largest value (78″–80″) is almost the same distance from the median as the smallest (60″–62″), and the frequencies are about the same for any two which are the same distance from the median. Such distributions are often found when we have a large number of cases to look at and when what we are measuring fluctuates randomly. In Chapter 8 we will discuss a very special example of a symmetrical distribution, called a normal distribution.

Now let's look at an example where the three measures do not have the same value. Wages or incomes make a good example because they are rarely distributed in an even or random fashion.

Example 13: Suppose there are 100 employees in a factory. They have the following distribution of salaries:

Class	Midpoint (x)	Frequency (f)	(fx)
$ 0– 9 thousand	$ 4.50	50	$ 225.00
10–19	14.50	24	348.00
20–29	24.50	6	147.00
30–39	34.50	5	172.50
40–49	44.50	4	178.00
50–59	54.50	4	218.00
60–69	64.50	3	193.00
70–79	74.50	2	149.00
80–89	84.50	1	84.50
90–99	94.50	1	94.50
			$1810 thousand

Mean: $\bar{x} = \dfrac{\Sigma\,(fx)}{n} = \dfrac{1,810,000}{100} = 18,100$

Median: 9,500

Mode: $4,500

Looking at the frequency histogram (Figure 2a) and frequency polygon (Figure 2b), we can see the very different character of this distribution compared to the one in Example 13.

Figure 2:

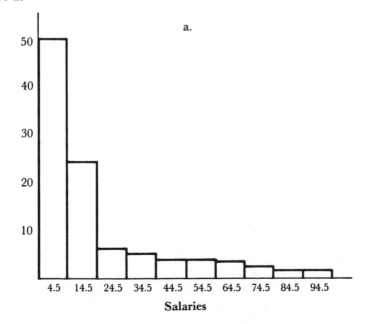

a.

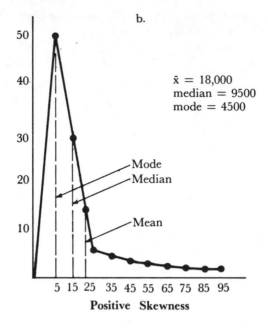

b.

$\bar{x} = 18,000$
median $= 9500$
mode $= 4500$

Mode
Median
Mean

Positive Skewness

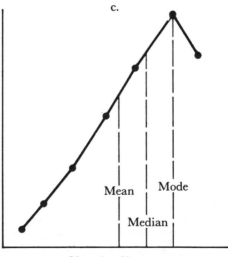

c.

Mean
Mode
Median

Negative Skewness

This distribution is highly **skewed;** that is, its highest point is to one side rather than in the middle. This pattern reflects the salary structure of the factory, where a very few people (presumably the owner and top executives) receive very high salaries, a slightly larger number (middle management) receive intermediate salaries, and the bulk of the employees (shop-floor workers) receive much lower ones. Notice that when this situation is graphed, the "tail" or low part of the graph is to the right, corresponding to the larger values; the mean is also to the right of the median and mode. We say that such a distribution is **skewed to the right,** or **positively skewed.** In the opposite case, where the trail is to the left and the mean is to the left of the median and mode, we say the distribution is **skewed to the left,** or **negatively skewed** (see Figure 2c).

Exercise: Can you think of a situation that would produce a negatively skewed distribution?

Whenever we have a skewed distribution, we get different information from the mean, median, and mode. In Example 13, the mean gives us a very misleading estimate of the average wage. The median gives us the salary of the person in the middle, who is probably a manual worker with some seniority and specialized skills. The mode gives the most common salary, probably that of a large group of younger, less skilled workers. Either the median or the mode — or maybe both — could be considered the "typical" wage, depending on our purpose.

The previous two examples illustrate some general characteristics of distributions. First of all we see that the mean, median and mode are similar only when we have a fairly regular or symmetric distribution. In that case the mean is the best measure, as it is in any distribution where the values are fairly close together and are not distorted by extremes. The reason the mean is the best measure under these circumstances is that it contains the greatest amount of information. In order to calculate the mean, we use the actual value of each item as well as the number of items. In calculating the median and mode, we really use only the number of scores and the value of one score. It is generally true that the more data that goes into a calculation, the more information we can get out of it. So whenever the amount of variation in a distribution is not too great or too uneven, we should use the mean.

Even in highly skewed distributions, like that in Example 13, the mean can give us important information if we ask the right questions. For example, suppose a firm decides to double the size of its operations. At the same time it wants the ratio of executives to workers to remain constant and has instructed its managers to calculate how much it would cost in additional wages to expand the company. How much would total wages be expected to rise using the mean as a guide? The median? Do you see why the mean is more accurate and reliable than other measures? In this case the mean will be a better guide than the median or mode because it has the advantage that it can always be multiplied by the total of the frequencies to get the total of the data points. The median, for example, cannot.

When looking at skewed distributions, it is best to calculate all three measures and note the relationships among them. If we have a distribution in which a few very high values are balanced against many low ones, the median will be lower than the mean and will be a fairer measure of the typical value of the distribution. If we have a distribution in which a few very low values are balanced against many high ones, the median will again be the fairest measure, though this time it will be higher than the mean.

The usefulness of the mode will depend on the extent of clustering in the distribution. In a very uneven distribution where the majority of cases are in one class, the mode is the most useful measure. When a distribution is found to be bimodal, this is often a useful piece of information. It may indicate that two different groups have been put together into one distribution or that there are two factors at work determining the values of the data. For instance, if a teacher grades a quiz during the first week of school and notes one mode in the high range and one in the low, it may indicate that some of the students have bought and read the assigned text while the other group hasn't gotten around to it yet.

Example 14: Test scores on a quiz:

Class	Midpoint	Frequency
0–9	4.5	1
10–19	14.5	2
20–29	24.5	1
30–39	34.5	8
40–49	44.5	2
50–59	54.5	1
60–69	64.5	3
70–79	74.5	3
80–89	84.5	8
90–99	94.5	1

Exercise: Find the mean, mode and median of the following distribution:

Class	Midpoint (x)	Frequency (f)
1–3	2	16
4–6	5	11
7–9	8	8
10-12	11	9
13-15	14	6

Is the distribution positively or negatively skewed? Why?

Now that we have some idea of correct uses of the mean, median and mode, let's look at some examples of how they are misused.

PART B: Analyzing Averages

The Inappropriate Average

Averages, like other statistical measures, provide ample opportunity for giving the wrong impression while sounding impressive. One area in which this is particularly true is in statements about the relative wealth or poverty of particular nations or groups of people. We have already shown that when there is an unequal distribution, the mean is not a fair measure of average or typical values. Yet people persist in talking about **per capita income.** Per capita income is simply the total income of the nation divided by its population — that is, the mean income per person. It is easy to sneer at foreign dictators who hide the vast wealth of a few and the incredible poverty of most by using such figures. But we need not go so far afield.

Example 15: In its June 30, 1975 issue, *US News & World Report* gave the following figure for US per capita income (that is, mean income per person) in 1974: $5,434. Median income is usually given per family. Since the 1975 *US Statistical Abstract* tells us that the average family size in 1974 was 3.44, the mean family income in 1974 was $18,692. But is that an indication of how much the average family made? Not at all. The median family income for 1974, again from the *Statistical Abstract,* was $12,836.

Dubious Averages

There are certain kinds of numbers which the rules of mathematics will not allow us to average because the result would be meaningless. Percents with different bases are a good example. Say the cost of living went up 8% last year and 12% this year. It may seem reasonable to say that on the average, the cost of living has gone up 10% in the last two years. But this is not true.

Example 16: According to the Bureau of Labor Statistics, the price of food increased by 12.2% from December 1973 to December 1974, and by 6.5% from December 1974 to December 1975. Suppose we bought $10 worth of groceries in '73. In '74 those groceries would cost
$$\$10 + (12.2\%)(\$10) = \$10 + \$1.22 = \$11.22$$
In December of 1975, the same groceries would cost
$$\$11.22 + (6.5\%)(\$11.22) = \$11.22 + \$.73 = \$11.95$$

The total increase for the two years is $1.95. Therefore the percent increase for the two years is:

$$\frac{\$1.95}{\$10} = 19.5\%$$

The average increase per year is:

$$\frac{19.5\%}{2} = 9.75\%$$

If we had simply averaged the two percents, we would have gotten

$$\frac{12.2\% + 6.5\%}{2} = \frac{18.7\%}{2} = 9.35\%, \text{ an error of } .4\%$$

If you encounter any difficulty in understanding this example, go back to Chapter 2 and review Example 1.

No Appropriate Average

Besides the cases where the wrong measure or the wrong method is used to find an average, there are many and varied examples of cases where people insist on quoting an average where none is appropriate. This is the kind of situation characterized by the classic joke about the statistician who sat with his head in the furnace, his feet in a bucket of ice water, and claimed that on the average he was quite comfortable. His mistake was to confuse an average with a standard. The following examples show some consequences of such reasoning.

Example 17: In her column in the *Boston Globe* on August 14, 1976, Ann Landers responded to a woman who complained about an insensitive doctor by stating: "Nothing shocks me anymore, especially when I know that 50 percent of the doctors who practice medicine graduated in the bottom half of their class." This would be no joke if the average graduate of a medical school just met the standard of adequacy. Then half our doctors would, by definition, be incompetent. Ideally, then, the average cannot be the standard — in this case it must be high enough so that the lowest member of the class still meets some basic standard of adequacy.

Industry also is likely to use averages and ignore variation. A favorite here is to cater to the mode and ignore the rest of the population. This may be good for business but unfortunate for the non-average consumer.

Example 18: *Parade Magazine* of June 13, 1976 came out with the following statement: "A survey ordered by the Motion Picture Association of America reveals that 80 percent of the over-60 population in this country never attends motion pictures.

The remaining 20 percent attends infrequently. There are 32 million Americans aged 60 and over. Thus, most films are made primarily for filmgoers in the 16–30 age bracket." What happened to the kids under 16, or the people between 30 and 60? It looks as though the population had been divided into 4 or 5 classes. The mode turns out to be the class of 16–30 year olds and the other age groups are ignored as far as the movie industry is concerned. If one were to do a frequency distribution on kinds of interests, the television industry would be even more guilty of this kind of thinking.

Comparing Averages

Comparing averages is always a dangerous procedure. Some pitfalls here include comparing averages that are based on data drawn from different sources, at different times, or with different amounts of variability.

Example 19: In *The Permissible Lie,* Samm Sinclair Baker gives the following example: "In another instance, two full-page TV ads appeared in the same issue of *The New York Times.* Both heralded results of national Nielsen reports. On page 94, NBC Television Network bragged in large type: 'THE BIG-GEST AVERAGE NIGHTTIME AUDIENCE.' On page 96, CBS television network boasted in bold type: '**THE BIGGEST AVERAGE NIGHTTIME AUDIENCE.**'

Again it was vital to read the smaller type (advertisers know that comparatively few people do). The NBC claim was 'according to the national Nielsen reports for the season to date.' The CBS contention was based on 'the latest national Nielsen report.' "

Comparing averages over time is particularly misleading where money is concerned. The value of the dollar can change drastically from year to year, so that an average income of $100 a week spelled wealth during the depression of the 30's, but falls below the poverty line for a family of four during the recession of the 70's.

Finally, we must be aware of the amount of variation in the data which is being averaged. It makes no sense to compare a city with a temperature range from 110° to −40° with one where temperatures range from 80° to 50° by saying that both have an average yearly temperature of 69°. Likewise, if one industrial plant puts out very high levels of pollutants in the summer and very low levels in the winter, we cannot compare its average to that of another plant with a steady intermediate level. Nor can we compare per capita incomes of nations with different income distributions. If the comparison between two averages is to be meaningful, the averages must arise from comparable data bases.

SUMMARY

PART A

An **average** is a number which represents a typical value of a set of numbers, or distribution. The most commonly used averages are the mean, median and mode.

The **mean** (or arithmetic mean, as it is technically called) is the sum of all the numbers in the distribution divided by the number of numbers. There are two formulas for the mean:

Formula 1: $\qquad\qquad \bar{x} = \dfrac{\Sigma\, x}{n}$ for ungrouped data

Formula 2: $\qquad\qquad \bar{x} = \dfrac{\Sigma\, (fx)}{n}$ for grouped data

The **median** is the middle score in a distribution if there is an odd number of scores, or the average of the two middle scores if there is an even number. If the data is grouped, the midpoint of the class containing the middle score (or the average of the midpoints of two classes, if necessary) is used.

The **mode** is the score with the highest frequency. For grouped data, the midpoint of the class with the highest frequency is used. If there are two such scores or classes, the distribution is called **bimodal.** If there are more than two, it is **multimodal.**

When the mean, median and mode are identical for a given distribution that distribution is **symmetrical.** If they are different, the distribution is **skewed.** If the lower part of the graph is to the right and the mean is larger than the median and mode, the distribution is said to be **positively skewed** or **skewed to the right.** If the lower part of the graph is to the left and the mean is smaller than the median and mode, the distribution is said to be **negatively skewed** or **skewed to the left.**

Each of the three averages has a different use. The mean contains the most information, but is misleading when the distribution is very skewed. The median gives a more typical value for skewed distributions. The mode is useful when there is much clustering in one or two classes. Bimodal distributions may indicate that two groups are included in one distribution.

PART B

Wrong impressions can be created by misusing averages in the following ways: using **inappropriate averages,** as for example using the mean of a highly skewed distribution as a typical value; averaging **numbers which cannot be averaged,** such as percents with different bases; using **averages as standards,** ignoring that values above the average or below will not meet that standard; **comparing averages** that cannot be compared because they are based on data from different sources, at different times, or with different amounts of variation.

EXERCISES

PART A

Use the following data for questions 1–4
   78, 89, 68, 95, 94, 62, 79, 89, 95, 68, 73, 82, 68, 74, 75

1. *Find the mean, median and mode of the above data.

2. Is the distribution positively or negatively skewed? Why?

3. *Make a frequency distribution for the above data using 4 classes.

4. Find the mean, median and mode of the grouped data.

5. *What can you say about any discrepancies that occur between #1 and #4?

6. Find some data that you are interested in averaging and find the mean, median and mode. (You'll need this for Part B.)

PART B

1. Explain the relative advantages or disadvantages of using the mean, median or mode in #6 above, and which position or argument would lead you to use which.

2. Explain what faults can be found with each of the following.
   a. *Parade Magazine* of November 16, 1975 quoted the following information from a study of American cities: "There was little difference in the average size of family between city and suburban residents in 1974 and no difference in the average number of children per family."
   b. A math class produced the following test scores:
      100, 100, 95, 95, 90, 85, 85, 80, 0, 0
   The teacher announced that the average score was 73.
   c. *A chemicals factory announces that its average pollution emission during the month was within the bounds of permissible levels.
   d. A magazine advertises that the median age of its readers is 34 and their mean salary is $25,000.
   e. *In order to earn a C a student must have an average of 75 or above. Joe's scores are 100, 74, 69, 65, 73. He gets a C.
   f. Last year the average salary increased by 5%. This year it increased by 6%. People's standard of living has increased this year over last year.

3. Look for examples of inappropriate averages used in advertising or reporting in the newspapers or magazines that you read. Try to analyze the faults in any questionable examples you find.

Chapter 7

VARIABILITY

PART A: Measures of Variability

In the last chapter we looked at measures which attempt to characterize a whole distribution with a single number. And we noted that although
these measures can be very helpful, they can also be misleading. In particular, we saw that one problem with averages was that although they might
tell us, with greater or less accuracy, the "typical" value of the distribution, they tell us nothing about other characteristics of the distribution,
such as how large it is, how spread out, or how compact. That is, they tell
us nothing about the amount of variability in the distribution. Since these
characteristics of a distribution affect the value or meaning of the average,
as well as being important in themselves, we now turn our attention to
measuring variability. To see the importance of these measures, look at the
following two sets of test scores.

Example 1: Scores: 3, 4, 4. 5, 5, 5, 5, 5, 6, 6, 7
              Mean: 5   Median: 5   Mode: 5

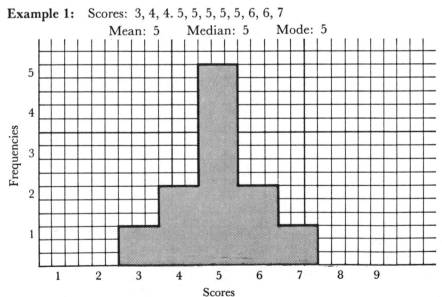

Example 2:   Scores: 1, 2, 2, 4, 5, 5, 5, 6, 7, 9, 9

Mean: 5   Median: 5   Mode: 5

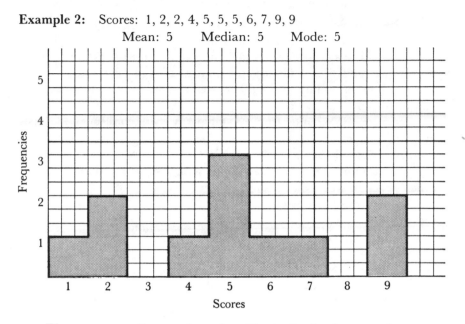

The means, medians and modes of both distributions are the same, yet the distributions present very different pictures. If we look at the graphs, we notice that one difference is how spread out the distributions are. The scores in Example 1 are all contained between the scores of 3 and 7, while those in Example 2 range from 1 to 9. Another difference is that the scores in Example 1 are bunched together in the middle, while those in Example 2 are more spread out toward the edges. Two measures will help us describe these differences: the range and the standard deviation.

The Range

The range tells us the distance from the largest to the smallest score. We find it by subtraction:

R = H − L where R is the range, H is the highest score and L the lowest. The range in Example 1 is

$$7 - 3 = 4$$

The range in Example 2 is

$$9 - 1 = 8$$

Notice that the range gives us only one limited bit of information. It tells us how far the highest score is from the lowest. It does not tell us anything about how the scores are distributed within that spread. For that we need a more complicated measure.

Exercise: Find the range for the following set of data: 16, 5, 11, 25, 12, 8, 4, 20

The Standard Deviation

The standard deviation is more complex than the range but is more useful as well. It tells us how all the scores — not just the extremes — are spread out in relation to the mean. One can think of the standard deviation as a kind of average of the differences between each individual score and the mean of all the scores.

Using the distribution from Example 1, we will go through the steps involved in calculating the standard deviation. Since we want to find the average of the differences between the individual scores and the mean, we begin by finding the individual differences first. These differences are called **deviations** from the mean. Since the mean of the distribution was 5, the deviations will look like this:

Score	Deviation
3	$3 - 5 = -2$
4	$4 - 5 = -1$
4	$4 - 5 = -1$
5	$5 - 5 = 0$
5	$5 - 5 = 0$
5	$5 - 5 = 0$
5	$5 - 5 = 0$
5	$5 - 5 = 0$
6	$6 - 5 = 1$
6	$6 - 5 = 1$
7	$7 - 5 = 2$

We find that we cannot average our deviations by simply adding them up and dividing because when we add them up we always get 0. This is not a peculiarity of this distribution, it is true of all distributions. If you think about it, you will see why. Since the mean is the arithmetic center of the distribution, all the deviations in one direction must cancel out all the deviations in the other direction, making their sum 0. We could just disregard the plus and minus signs and use the absolute values, or distances, to find the average. But it turns out that we get a much more useful measure if we solve the problem by squaring each deviation. That is, we multiply each deviation by itself. We know that two negative numbers multiplied together give a positive product, and of course two positive numbers do also. So if we square the deviations, we will have a positive sum to work with, which we can then divide by the number of scores in the distribution to get our average. Continuing our example:

Score	Deviation	Deviation Squared
3	−2	4
4	−1	1
4	−1	1
5	0	0
5	0	0
5	0	0
5	0	0
5	0	0
6	−1	1
6	−1	1
7	−2	4
		12

The sum of the squared deviations is 12; there are 11 scores in our distribution, so the average of the squared deviations is

$$12 \div 11 = 1.0909$$

This average of the squared deviations is called the **variance.** The variance is an important measure in scientific statistics, but not in economic or social statistics. Since we will not be concerned with any of the special applications, such as analysis of variance, we will not be using it here. Instead, we will continue to the last step in our calculation of the standard deviation. The standard deviation is the square root of the variance. Its advantage is that it measures the distance between the average score and the mean in the same units as the original score — not in squares of that unit. For example, if we were working with a distribution of people's heights, measured in inches, the variance of the distribution would be in square inches. The standard deviation, being the square root of the variance, would be in inches, just like the individual heights. This makes it much easier for us to interpret. The standard deviation of our present distribution, then, is

$$\sqrt{1.0909} = 1.04$$

The easiest way to find a square root is to use a pocket calculator. Next best is to use a table of square roots. A table and instructions for use are provided in Appendix Table 1.

We can write a formula for the standard deviation using the same symbols we used to write the formula for the mean:

Formula 1: $$\sigma = \sqrt{\frac{\Sigma (x - \bar{x})^2}{n}}$$

We can translate this formula into the following English sentence: The standard deviation is equal to the square root of the average of the squared deviations. A more useful way to view it is as a reminder of all the steps required to calculate the standard deviation. Remember that x represents each individual score, $\bar{x}$ the mean of the scores, n the number of scores in the distribution, and Σ the operation of addition. We will use σ (the lower case Greek letter sigma) to represent the standard deviation of a population. We know from experience with algebra, that the place to start is within the parentheses. The other steps follow more or less naturally:

1. Find the deviations by subtracting the mean from each score in turn.
2. Square each deviation.
3. Add up the squared deviations.
4. Divide that sum by the number of scores in the distribution.
5. Take the square root of the average of the squared deviations.

If the data we are dealing with is grouped, then there is a minor change in the formula and in the procedure which it represents:

Formula 2: $$\sigma = \sqrt{\frac{\Sigma f (x - \bar{x})^2}{n}}$$

The x, instead of standing for individual scores as in Formula 1, now stands for the midpoints of the classes, and the f stands for the frequencies. The steps represented by this formula are:

1. Find the deviations by subtracting the mean from the midpoint of each class.
2. Square each deviation.
3. Multiply each squared deviation by the frequency for that class.
4. Add up the products of the squared deviations and frequencies.
5. Divide that sum by the number of scores in the distribution (sum of frequencies).
6. Take the square root of the average of the squared deviations.

Before going through the 6 steps, we group the data in Example 2 and find the mean.

Class	Class Midpoint	Frequency
1–3	2	3
4–6	5	5
7–9	8	3

The mean for the grouped data in this case turns out to be the same as for the ungrouped data:

$$\frac{(3)(2) + (5)(5) + (3)(8)}{11} = \frac{55}{11} = 5$$

Now we can go through the 6 steps described by formula 2.

1. $(x - \bar{x})$

Midpoint		Mean		Deviation
2	−	5	=	−3
5	−	5	=	0
8	−	5	=	3

2. $(x - \bar{x})^2$

Deviation	Deviation Squared
−3	9
0	0
3	9

3. $f(x - \bar{x})^2$

Deviation Squared	Frequency	Frequency times Deviation Squared
9	3	27
0	5	0
9	3	27

4. $\Sigma f(x - \bar{x})^2$

   $27 + 0 + 27 = 54$

5. $\dfrac{\Sigma f(x - \bar{x})^2}{n}$

   $54 \div 11 = 4.9091$

6. $\sqrt{\dfrac{\Sigma f(x - \bar{x})^2}{n}}$

   $\sqrt{4.9091} = 2.22$

Now that we have gone through all the steps and learned how to use the above formula, we must admit that this formula does not provide the quickest or easiest way to find the standard deviation. The reason we introduced the formula in this form is that it allows us to see clearly what standard deviations are all about. We see that we are averaging the squared deviations and then taking the square root. This gives us a measure of variability, of how far the average score is from the mean, or how scattered the data in the distribution actually are. We have calculated two standard deviations. We found that for the distribution in Example 1, the standard deviation was 1.04, and for the distribution in Example 2 it was 2.22, or more than twice as large. This should not surprise us, since the distribution in Example 2 is much more scattered or dispersed than the distribution in Example 1.

Exercise: Calculate the standard deviation using Formula 2 for the following data:

Class	Class Midpoint	Frequency
2–4	3	3
5–7	6	5
8–10	9	7
11–13	12	4

Although Formula 2 helps to make the meaning of standard deviations clear, it is a nuisance to calculate. Subtracting each score from the mean was tedious enough when we had fewer than a dozen scores to work with. Imagine using this technique for hundreds of numbers! When a large number of scores is involved, the formula below is much easier to use. (Even this formula, though, generally requires use of a computer, or at least a calculator, if there's a large amount of data.)

Formula 3:
$$\sigma = \sqrt{\frac{\Sigma fx^2}{n} - \left(\frac{\Sigma fx}{n}\right)^2}$$

This formula is mathematically identical to the one we used earlier and gives exactly the same results. It is easier to remember if you think of it this way:

The standard deviation is the square root of the difference between the mean of the squares and the square of the mean.

Using the data from Example 2, we can construct the following table:

Midpoint (x)	Midpoint Squared (x^2)	Frequency (f)	(fx)	(fx^2)
2	4	3	6	12
5	25	5	25	125
8	64	3	24	192
		11	55	329

Now let's go through the steps of Formula 3. Here we can begin either with the mean of the squares or the square of the mean. Since we usually want to calculate the mean before doing the standard deviation anyway, we can start there.

1. $\left(\dfrac{\Sigma fx}{n}\right)^2$ $(55 \div 11)^2 = 5^2 = 25$

2. $\dfrac{\Sigma fx^2}{n}$ $329 \div 11 = 29.909$

3. $\dfrac{\Sigma fx^2}{n} - \left(\dfrac{\Sigma fx}{n}\right)^2$ $29.9091 - 25 = 4.9091$

4. $\sqrt{\dfrac{\Sigma fx^2}{n} - \left(\dfrac{\Sigma fx}{n}\right)^2}$ $\sqrt{4.9091} = 2.22$

Exercise: Find σ using Formula 3 for the data below.

Class	Midpoint	Frequency
1–5	3	2
6–10	8	6
11–15	13	2

The formulas given in this chapter so far are used to find the standard deviation of a whole population when all the data from that population is available. More often, however, we only look at a sample which we believe represents the population. When that is the case, we modify the formula slightly. For reasons which we need not pursue here, the results are somewhat more accurate if we use $n - 1$ instead of n in the denominator. We use s for the standard deviation of a sample instead of σ, which we use only for the standard deviation of a whole population. The formula is:

Formula 4:
$$s = \sqrt{\frac{\Sigma f (x - \bar{x})^2}{n - 1}}$$

Exercise: Using Formula 4, find the standard deviation of the following distribution:

Class	Frequency
10–14	4
15–19	6
20–24	2
25–29	8
30–34	3

The practical usefulness of the standard deviation will gradually become clearer in the next few chapters. For now, it is enough to realize that it gives us a way of judging how scattered a distribution is, that is, the dispersion around the mean. Although the actual number we calculate tells us very little by itself, it should be obvious that if two distributions are expressed in the same units (inches) and have the same mean, as those in Examples 1 and 2, the one with more scattering will have the larger standard deviation. Of course we do not usually find distributions with exactly the same means, and often we want to compare distributions measured in different units. We can overcome these difficulties with another measure.

Coefficient of Variation

The coefficient of variation is a fancy name for a simple measure: the standard deviation divided by the mean.

$$CV = \frac{s}{\bar{x}}$$

The advantage of this measure is that it gives us a relative measure of variability — that is, the amount of variability relative to the size of the mean. Therefore we can use it to compare distributions with different means. We can also compare distributions measured in different units, since dividing a deviation measured in, say, inches by a mean which is also measured in inches, results in canceling out the inches. Coefficients of variation are always expressed by converting the result of division into a percent.

The advantage of this measure is that since the standard deviation and the mean are always expressed in the same units, the division cancels out these units and gives us a relative measure of variability, that is the amount of variability relative to the size of the mean.

Example 3: Suppose we want to compare variation in performance on two tests. One test is scored according to the number of problems done correctly. It has a mean of 6.5 and a standard deviation of 2. The other test is scored on a percent basis and has a mean of 60% and a standard deviation of 15%. Which test distribution is more scattered — shows greater variability? For the first test, the coefficient of variation is

$$CV = \frac{2}{6.5} = .31$$

For the second test we have

$$CV = \frac{15\%}{60\%} = .25$$

This calculation allows us to see that there is more variability in the distribution of the first test scores than the second.

Notice that the coefficient of variation is a ratio which tells us how the standard deviation compares to the mean. If it is a small fraction of the mean, there is a small amount of scattering, or variation. If it is a large fraction of the mean, or, in rare cases, even larger than the mean, then there is a great deal of variation in the distribution. In the next chapter we will see another way in which standard deviations can help us describe distributions. Comparing the two measures of variability, s always refers to absolute variability and CV to relative variability.

Exercise: Find the CV of the previous exercise.

PART B: Vagaries of Variability

The main way in which measures of variability are misused is that they are ignored. Sometimes we are presented with statistics which fail to even mention the problem of variation. Sometimes a standard deviation or range is quoted but its implications are ignored by author and reader alike.

Too Much Variability

We have already pointed out in Chapter 5 that if there is too much variation, the average is not a useful measure. Range and standard deviation are not the only ways to measure variability.

Example 4:   The September 26, 1976 *Boston Globe* ran a story on income distribution. It said, in part,

"The number of people living in what the government officially defines as poverty ($5500 for a family of four) increased by 2.5 million or 10.7 percent in 1975, the Census Bureau reported yesterday. . . .

The total below the poverty line rose to 25.88 million persons, about one American in eight. . . .

The census agency said median family income rose to $13,719 last year, up $817 or 6.3 percent from the median in 1974."

Notice that we do not need to see the standard deviations to become aware of the variability of incomes. Variability is indicated by the difference between the number of people below the poverty line and the median family income. While the median income in 1975 was $13,719, one eighth of the American population still lived substantially below the poverty line (defined as $5500 for a family of four). The median quoted by itself makes it sound as though the "average" American were quite comfortable. But the median income only tells us about the middle of the distribution. To make sense out of the numbers above, we would need some measure of variability to tell us about the "non-average" Americans.

Another case in which we must be concerned about high variability is one in which we want to use variability as a measure of reliability. For example, the durability of a car can be measured, as some advertisers have done, by its mean life expectancy. But how reliable is this measure? We don't know if we don't know the standard deviation as well as the mean.

Example 5:   Consider the following two cars:
Car A. Mean life expectancy: $\bar{x} = 6$ years, $s = 1$, $CV = 1/6 = 16.67\%$
Car B. Mean life expectancy: $\bar{x} = 7$ years, $s = 3$, $CV = 3/7 = 42.85\%$

Which should you buy if you want a car that will last at least 5 years? By relying on the information above, you would select Car A. In this case, we can see intuitively that Car B promises a slightly longer life but the promise is much less reliable. We will return to this example in the next chapter and show how to make the comparison more exactly.

Too Little Variability

Sometimes it is to the advantage of the advertiser not to inform potential customers that there is actually very little variation among different manufacturers' products. Here is a case in point.

Example 6:   The following list shows the amount of protein contained in ten different breakfast cereals.

Corn Flakes	2 grams
Rice Krispies	2
Shredded Wheat	2
All Bran	3
Product 19	3
Total	3
Wheaties	3
Cheerios	4
Post Fortified Oat Flakes	5
Special K	6

Special K is advertised as a high protein cereal. Its protein content is twice that of the group median (3). But does it make any difference? The minimum protein required per day by the average adult is approximately 45 grams. How much difference will it make in one's overall nutrition if 6 rather than 3 of those grams are contained in the breakfast food?

Comparisons

In Part A, we mentioned that when comparing scores from distributions with different variabilities, the standard deviation must be taken into account. Let's look at a specific example.

Example 7:   An article in *Science* of June 10, 1977 describes an experiment in which some children were given vestibular stimulation (they were spun around in special chairs) to see if this would improve their motor development. There were three groups of children, the T group which got spun, the CH group which got held in the chair but not spun, and the CNH group to which nothing was done. Below are the motor skills test scores for all three groups at the end of the experiment.

Group	Mean Score	Standard Deviation
T	86.38	35.66
CH	77.83	35.10
CNH	60.50	39.94

Note that the group that got the treatment scored highest on the average. Were there people in the T group that scored lower than people in the CH group? In the CNH group? We don't know how many, but there must have been some. How do we know?

Finally, large standard deviations make comparisons very tricky, not only when we are comparing groups, but even more so when we are comparing an individual to a group. Look at Example 8.

Example 8:   The following distribution contains eight test scores. Assume that 60 is a passing mark.

$$R = 80$$

80, 70, 60, 60, 60, 50, 0, 0 $\bar{X} = 47.5$

$$s = 30.59$$

The person who scored 50 failed the test, but did better than the mean.

SUMMARY

PART A

In this chapter we have discussed the importance of variability and have introduced three measures: the range, the standard deviation and the coefficient of variation. We also mentioned the **variance,** which is the square of the standard deviation.

The **range** is the difference between the smallest and the largest data point in the distribution.

R = H − L where R is the range, H is the highest score or largest data point, and L is the lowest or smallest.

The **standard deviation** uses all the data points to find how far they are, on the average, from the mean. There are several equations for calculating the standard deviation. Formulas 1, 2, 3, refer to the population standard deviation (σ) and Formula 4 (s) to the sample standard deviation.

Formula 1: $\sigma = \sqrt{\dfrac{\Sigma\,(x - \bar{x})^2}{n}}$ ungrouped data

Formula 2: $\sigma = \sqrt{\dfrac{\Sigma f\,(x - \bar{x})^2}{n}}$ grouped data

Formula 3: $\sigma = \sqrt{\dfrac{\Sigma fx^2}{n} - \left(\dfrac{\Sigma fx}{n}\right)^2}$ grouped data (calculation formula)

Formula 4: $s = \sqrt{\dfrac{\Sigma f\,(x - \bar{x})^2}{n - 1}}$ grouped data (when the standard deviation is calculated from a sample rather than from a population)

The **coefficient of variation** is used to compare distributions with different means and different standard deviations. The larger the value of the coefficient, the more variation there is in the distribution.

$$CV = \frac{s}{\bar{x}}$$

PART B

The main error involving measures of variability is neglect. It is important to consider them in the following cases.

> **Too much variability** makes averages meaningless and implies less reliability.
> **Too little variability** makes comparisons irrelevant.
> **Comparisons** between distributions and comparisons of one data point to the mean of a distribution must always take variability into account.

EXERCISES

PART A

1. Write Formula 3 for calculating the standard deviation on a piece of paper. Underneath, write all the steps you would take to do that calculation for ungrouped data. Then write all the steps for grouped data. Compare your steps with those in the text (p. 138).

2. *By inspection only decide which of the following distributions you would expect to have the larger standard deviation.
   Distribution A: 2,   5,   7,   9,   10,   13,   16,   18,   20,   25
   Distribution B: 10,   11,   11,   12,   12,   12,   13,   14,   14,   15

3. Calculate the standard deviation for the two distributions in #2 and see if your guess was right. Use Formula 1 to calculate the standard deviation of Distribution A and Formula 3 for Distribution B.

4. Below are some of the test scores from a statistics class. Do a frequency distribution, find the mean, median, mode and standard deviation.
   35, 35, 35, 40, 40, 50, 50, 55, 55, 55, 60, 60, 60, 60, 65, 70, 70, 70, 70, 70, 70, 75, 75, 75, 75, 90, 95

5. *Suppose the people in one neighborhood have a mean income of $6,000 with a standard deviation of $1,500 and those of another have a mean income of $25,000 with a standard deviation of $5,000. Which neighborhood shows more variability of income? Why?

PART B

1. Give some examples of averages you have come across in your reading that you would like to know the standard deviations for. Explain why you would like to know them and what you would expect them to look like.

2. Consider Example 5 again. Suppose you wanted to buy the car that offered you the best chance of lasting 15 years. Which would you pick? How good would you estimate your chances were of your choice actually lasting that long?

3. *Calculate the mean and the standard deviation for the data below:

    90, 75, 70, 60, 55, 50, 25, 15, 10, 0, 0, 0, 0

    Give two reasons why the mean is a poor measure of the typical value of the distribution. Is the median better? The mode? In what way are they better and in what way are they not?

4. Here is more data from the experiment quoted in Example 7. On a reflex test, the scores of the three groups were:

Group	Mean Score	Standard Deviation
T	62.54	5.64
CH	58.00	9.73
CNH	53.93	7.76

    The percent of increase in score of the T group over the CH group for the motor skills test was:

    $$86.38 - 77.83 = 8.55; \quad \frac{8.55}{77.83} = 11\%$$

    Calculate the percent of increase of the T group over the CH group for the reflex test. Compare the two. Which is greater? Are you more convinced by the increase on the motor skills test or the reflex test? Why?

5. *Two people, Jim and Jane, were tested on two different tests, each scored on a scale from 0 to 100. Each scored 10 points above the mean. The CV on Jim's test was .65; the CV on Jane's was .4. Who did better? Why?

Chapter 8

NORMAL DISTRIBUTIONS

In the past several chapters, we have used histograms to picture particular distributions, and have calculated the mean, median, mode, range and standard deviation to describe their most important characteristics. On each occasion, we have been dealing with specific examples of distributions. Now we are going to look at an ideal rather than a real distribution. This ideal, which statisticians call "the normal distribution," is best thought of as a mathematical model rather than an actuality. This point becomes clearer when we examine a more familiar example from geometry.

In geometry we talk about perfect circles. These perfect circles have characteristics never found in the real world: each point on the circumference is *exactly* the same distance from the center as each other point; the line forming the circumference has length but no width, etc. Such characteristics are impossible to duplicate in the real world, but by assuming them for the perfect circle we make it easier to talk about circles in general. We can then use concepts and formulas based on our study of perfect circles for such practical applications as calculating a spaceship's orbit or figuring the area of a dime. The perfect circle we talk about in geometry, while it doesn't exist in the real world, makes a good mathematical model for the round objects that do exist and lets us perform calculations about them.

In the same way, the so called normal distribution is a mathematical model. There is nothing "normal" about it; in fact it would be highly abnormal to find a perfect example of it in real life. But it provides us with a mathematical way of dealing with all distributions that more or less resemble it. We know that the more nearly a real object resembles the perfect circle, the more accurate our calculations of its area or circumference will be. Similarly, the more nearly a distribution approaches the characteristics of a normal distribution, the more accurate our calculations based on those characteristics will be. But before we discuss how well the characteristics of a normal distribution fit any given distribution, we must determine what those characteristics are.

PART A: The Normal Distribution and the Z Score

The Normal Distribution

It will be easier to understand the characteristics of a normal distribution if we look at its graph. A typical normal distribution looks like this:

Figure 1

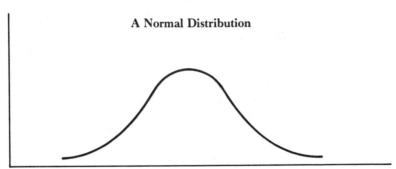

A Normal Distribution

The curve is the result of a very regular kind of frequency distribution made up of a large number of scores and a large number of classes. Consider the following example.

Example 1:   Suppose we have a bag which contains 50 pennies. We dump the bag out on the floor and count the number of heads we see. Then we put all the pennies back in the bag and repeat the process. Say we repeat it about 400 times. When we're through, we make a frequency distribution of the number of times we saw 10 heads, 11 heads, 12 heads, and so on. (It's unlikely that out of 50 pennies fewer than 10 or 12 heads would appear.) Suppose now that our frequency distribution looked like this:

Number of heads	Frequency	Number of heads	Frequency
11	1	25	33
12	2	26	32
13	3	27	29
14	4	28	24
15	6	29	21
16	7	30	17
17	10	31	13
18	12	32	9
19	16	33	7
20	21	34	5
21	25	35	4
22	29	36	3
23	31	37	2
24	33	38	1

We could group these 28 outcomes in a number of ways. Since 28 is evenly divisible by four and by seven, two obvious possibilities are to use four classes or to use seven classes.

Number of heads	Frequency	Number of heads	Frequency
11–17	33	11–14	10
18–24	167	15–18	35
25–31	169	19–22	91
32–38	31	23–26	129
	400	27–30	91
		31–34	34
		35–38	10
			400

Below we can see the three histograms corresponding to the two grouped (Figures 2a and 2b) and the ungrouped (Figure 2c) versions of the data. Notice that the larger the number of classes used, the more the histogram looks like a normal curve. The ungrouped version, which has the largest number of classes (28 to be exact) looks most like the normal curve.

Figure 2a

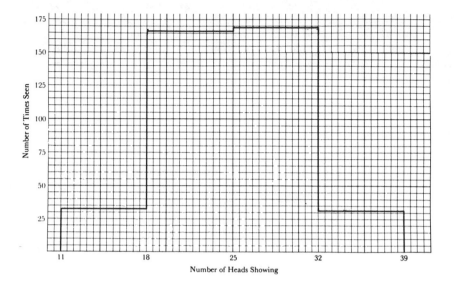

Figure 2b

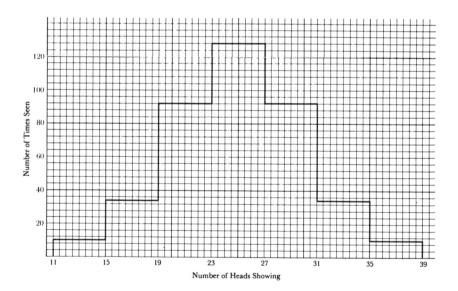

Figure 2c

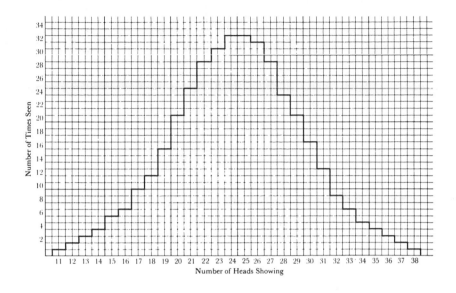

Imagine now that we put more and more pennies in the bag, increasing the number of classes. We also throw the pennies on the floor more and more often, thus minimizing accidental variations. Do you see that eventually we will get something that very closely approximates the mathematically perfect normal curve? Tossing pennies is one way to approximate a normal curve. There are many others, and we will talk about some of these in Part B. Now we can describe the characteristics of the normal curve.

When we say "the normal curve" we are not referring to a single shape. Graphs of normal distributions can assume many shapes, but all of these curves have certain properties in common; in this sense they are more like triangles and rectangles than they are like circles. All three curves in Figure 3 are normal curves.

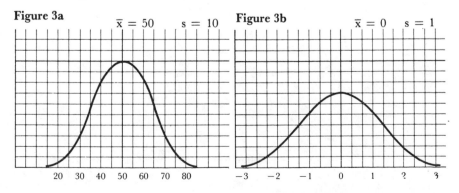

Figure 3a $\bar{x} = 50$ $s = 10$ **Figure 3b** $\bar{x} = 0$ $s = 1$

$\bar{x} = -2$ $s = 1$ **Figure 3c**

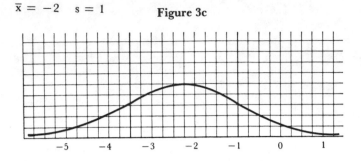

The characteristics of the normal curve are:

1. The midpoint of the curve is also the highest point, and the curve forms a bell shape. The ends of the bell theoretically extend infinitely in either direction, getting closer and closer to the horizontal axis but never quite touching it. (If we were tossing an infinite number of pennies, for instance, all conceivable numbers of heads would occur at least once, so no data point would ever have the frequency of 0.)

2. The curve is symmetric. This is just like the symmetrical histograms and polygons we discussed in Chapter 6. The distribution is not at all skewed: high values occur just as frequently as low values. The mean is therefore in the middle, corresponding exactly to the median. Since this mid-point is also the point with the highest frequency, it represents the mode as well. Since the mid-point represents the mean, median, and mode, 50% of the scores are to its right.

3. The curve is spread out in a particular way. Whatever the standard deviation may be, 68% of the area under the curve will be within that distance of the mean, 95% of the area will be within twice that distance, and 99.7% will be within three times that distance. Figure 4 shows this characteristic for a curve whose mean is 70 and whose standard deviation is 20.

Figure 4

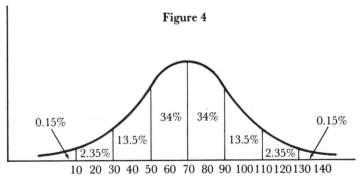

Note that 34% of the area is within 20 points on each side of the mean. These 20 points equal one standard deviation so we speak of 68% of the area (34% + 34%) as being "within one standard deviation of the mean." Likewise another 13.5% of the area is in each of the two parts of the graph between 20 and 40 points away from the mean; we say that 95% of the area (13.5% + 34% + 34% + 13.5%) is "within two standard deviations of the mean." And so on.

Although it may not be obvious at first sight, all these properties make it possible to define the curve of any normal distribution by stating the mean and the standard deviation. Here again the normal curve is like a circle: if we know the center and radius of the circle, we know where all the points on the circumference must be. If we know the center (mean) and the spread (standard deviation) of a normal curve, we know where all the points on the curve will fall.

Well, so what? First remember that in Chapter 5, we learned that the area of any section of a histogram or polygon is the same as its relative frequency. The same thing is true of a curve — finding the area under the curve between any two data points is the same thing as finding the relative frequency of the class bounded by those two points. Once we know the mean and the standard deviation of a normal distribution, we can calculate relative frequencies *without going back to the original raw data!* Notice that the vertical scales on all graphs of normal distributions presented so far have been left blank. This is because we do not need such a scale. Normal distributions are relevant only when we are dealing with very large numbers of data points or scores. As we noted in Chapter 5, actual frequencies are not as useful as relative frequencies when we are dealing with large distributions. Since we figure the relative frequencies by the area under the curve, we do not need a vertical scale for the actual frequencies.

This may appear to be some magical sleight of hand accomplished through the use of graphs, but it is not. Suppose we have a normal distribution of heights in a large population of people. We have calculated that the mean height is 65 inches and the standard deviation is 5 inches. This will reflect the fact that 34% of the people's heights are between 65 and 71 inches and another 34% are between 60 and 65 inches. That is a piece of knowledge which we can use without having to go back to masses of raw data to calculate.

Example 2:   Assume that a test has been given to thousands of students. The mean, median and mode of the percent scores are all in the neighborhood of 66%, and the standard deviation is approximately 11%. We can group the scores into classes which are one standard deviation (11%) in size, and graph the distribution like this:

Figure 5

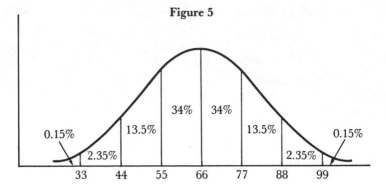

This graph allows us to see approximately how many students fall into each of the classes. If we are told that 10,000 students took the test, we can state with a fair degree of certainty that about 3,400 (34% × 10,000) students scored between 66% and 77%, while no more than 15 students (10.15% × 10,000) scored 100%.

Exercise: Assume that 5000 students took a statistics test. The mean, median and mode were all approximately 73% and the standard deviation about 8%. How many students scored between 73% and 81%? Between 57% and 65%?

Binomial Distributions. One very important example of a normal distribution is a binomial distribution. The distribution in Example 1 is a **binomial distribution.** That is, it shows the probability of each possible outcome in a situation involving two mutually exclusive alternatives — in this case a penny landing either heads or tails. Depending on how many times the two alternatives are tested, we may have different numbers of possible outcomes. If we tossed one penny only two outcomes are possible (either one head or one tail). Tossing 10 pennies gives 11 possible outcomes (anywhere from 0 to 10 heads may be showing). The binomial distribution shows the probability of each possible outcome.

A binomial distribution is approximated by the normal distribution whenever a reasonably large number of cases is considered and when the probability of the outcome we are looking at is not too close to the extremes (a sure thing or something that never happens). Therefore if we know the mean and the standard deviation of the distribution, we can calculate how likely any particular outcome we are interested in will be. The formulas for finding the mean and the standard deviation of any particular binomial distribution depend on two factors: the number of items in the sample (as for example the number of pennies being tossed or the number of voters being questioned) and the probability that the outcome will be one of the two possibilities (as for example the chance that a penny will come up heads or that a voter will vote for Candidate Jones). As always, we use n for the number of items in the sample. We will use the letter p for the probability of a particular outcome occuring. Probabilities, which we will take up in more detail in Chapter 10, are always expressed as numbers between 0 and 1, where 0 means there is no chance of something happening and 1 means it is a certainty. The probability of a penny landing heads is .5 since it is equally likely to land heads (50%) or tails (50%). The formulas for the mean and standard deviation of a binomial distribution are as follows:

1. Mean: $\bar{x} = np$ where $\bar{x}$ is the mean, n the number of items in the sample and p the probability of the desired outcome.

2. Standard deviation: $s = \sqrt{np(1-p)}$

Example 3: Suppose that we are selecting a sample of 36 people, $(n=36)$. We know that the probability of any given member of the sample being a woman is 50% or $p = .5$. What is the probability that our sample will include 12 or fewer women?

First we draw a graph of the distribution, using the formulas for the mean and the standard deviation:

$$\bar{x} = (36)(.5) = 18$$
$$s = \sqrt{(36)(.5)(.5)} = 3$$

Figure 6

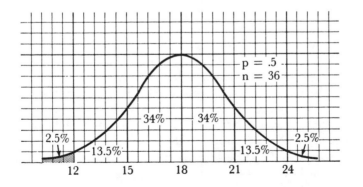

A quick glance at the distribution tells us that the probability that there will be 12 or fewer women in the sample is 2.5%.

Example 3 should make it clear why the mean of the binomial distribution is the size of the sample times the probability of the desired outcome. If we take a sample of 36 people and half of them are expected to be women, then the average sample should have half of 36 or 18 women in it. The justification of the formula for the standard deviation would involve us in too long a discussion, so we shall simply accept it as a given.

The Standard Normal Distribution and the Z-Score

Examples 2 and 3 were simple because the classes we wanted to know about were exactly one standard deviation in size. In other cases, more complex methods must be used, and the next few pages are devoted to explaining those methods. But already you may be wondering why it's necessary to learn any new methods of calculating relative frequencies, when we already learned perfectly adequate methods in Chapter 5. The answer is two-fold.

First, with very large amounts of data it is both time-consuming and expensive to go back to the raw data each time we have a new question. This is particularly true in the examples we will use in this chapter, most of which are calculations about scores on standardized tests given to large numbers of people.

But the second reason, which will concern us in many of the later chapters, is more important and is actually the reason that these methods were originally developed. It has to do with the making of projections (well-grounded guesses) in inferential statistics. Up to now we have been dealing with descriptive statistics — how to take data which exists in raw form and summarize it numerically and graphically. But in the remaining chapters we'll be concerned with inferential statistics — how to make guesses about a large population from a small sample. The methods we are using here to infer information we already have but don't want to bother to calculate exactly, we will use later on to infer information that we *don't* already have in raw form.

In Example 2 we used the idea of the normal distribution to find the relative frequency of a class whose boundary points (66% and 77%) were separated by one standard deviation (11%). Suppose we had wanted to know the relative frequency of students scoring between 44% and 73%? For this purpose we use a slightly different form of graph, and a new measure.

In Example 2, the numbers on the horizontal scale were actual scores, with the mean score in the center and the scores representing one, two and three standard deviations from the mean in either direction labeling the other points. We could just as well label the horizontal scale with the number of standard deviations each point is from the mean. Then the mean itself is represented by 0 (since it is no distance from itself) and the other points are 1, 2, 3, and −1, −2, −3 respectively.

Example 4:

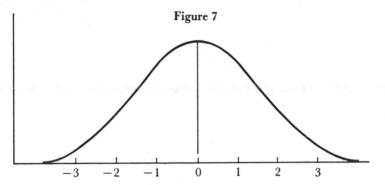

Figure 7

When we use this method of representing a normal distribution, we call it the standard form of the normal distribution. The standard form always looks exactly the same. If we redraw the three normal distributions in Figure 3 on the same size graph paper, using 0 as the mean and 1 as the size of one standard deviation, all three will look exactly like Figure 7 above. That is why it is called the **standard** normal distribution.

The **standard score** or **z-score** is a number which tells us exactly how many standard deviations (or fractions of a standard deviation) measure the distance between any selected score and the mean. Each actual score in a distribution has a corresponding z-score. In Example 2, the z-score for 66% is 0, since 66% is the mean and there is no (0) distance between the mean and itself. The z-score for 99% is 3, since the distance between the mean and 99% is 3 standard deviations ($3 \times 11\% + 66\% = 99\%$).

Example 5:   Figure 8 shows a normal distribution with a mean of 165 and a standard deviation of 12. It is marked with the actual scores along the horizontal scale and the standard scores, or z-scores, directly beneath them.

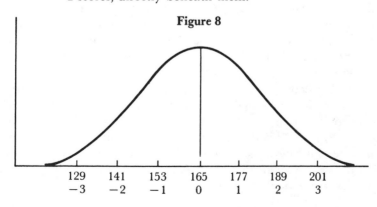

Figure 8

The z-score is the number of standard deviations the actual score is from the mean, or, to put it another way, it represents the distance between the actual score and the mean, divided by the standard deviation. The formula for z-scores is:

$$z = \frac{x - \bar{x}}{s}$$

where z is the value of the z-score or standard score, x is the actual score we are interested in, $\bar{x}$ is the mean, and s is the standard deviation.

Example 6: In a distribution with a mean of 165 and a standard deviation of 12, what is the z-score corresponding to the score of 160? Since $x = 160$, $\bar{x} = 165$ and $s = 12$, we have:

$$z = \frac{160 - 165}{12} = \frac{-5}{12} = -.42$$

The z-score is $-.42$

We interpret this z-score as meaning that the corresponding score of 160 is less than half a standard deviation (or .42 standard deviations to be exact) to the left of the mean. Notice that negative z-scores are to the left and positive z-scores are to the right of the mean.

Exercise: If a distribution has a mean of 85 and a standard deviation of 16, what is the z-score corresponding to 70?

Once we have a way of calculating z-scores, we can find the relative frequency of any class of scores we are interested in. Finding the z-scores of the two boundary points of any class is the first step to finding the area under the curve that represents that class. That area, as we explained in Chapter 5, is exactly the same as the relative frequency. In the case of a histogram, the area is easy to figure from the graph. Since the normal distribution is a curve, it would be impossible to get an accurate answer from the graph. There is a mathematical formula, but it is much too complicated for efficient use. Therefore a table has been constructed from the formula on which we can look up the areas we need. But we must use z-scores instead of the actual raw scores to use the table. Otherwise we would need infinitely many tables, one for each possible mean and standard deviation combination. The z-score allows us to get directly at that aspect of the normal distribution which determines the areas of each class: the relationship between the mean and the standard deviation. Using the standard normal distribution allows us to translate each of the possible different scores into one scale. Look again at Figure 3. A score of 60 in distribution 3a, 1 in 3b and −1 in 3c all have the same z-score of 1, since each is 1 standard deviation to the right of the mean. Therefore the z-scores allow us to use one table to look up areas corresponding to scores from all possible normal distributions.

There are several different kinds of tables that are used for this task. A very simple one is reproduced below. On this table, z-scores are given to only one decimal place. The number under the heading "area" gives the fraction of the total area which lies between the given z-score and the mean. A more complete table is given in the Appendix (p. 305).

z-score	area	z-score	area	z-score	area
0.0	.0000	1.0	.3413	2.0	.4772
0.1	.0398	1.1	.3643	2.1	.4821
0.2	.0793	1.2	.3849	2.2	.4861
0.3	.1179	1.3	.4032	2.3	.4893
0.4	.1554	1.4	.4192	2.4	.4918
0.5	.1915	1.5	.4332	2.5	.4938
0.6	.2257	1.6	.4452	2.6	.4953
0.7	.2580	1.7	.4554	2.7	.4965
0.8	.2881	1.8	.4641	2.8	.4974
0.9	.3159	1.9	.4713	2.9	.4981
				3.0	.4987

By telling us what percent of the total area under the curve lies between any given z-score and the mean, the table can be used to find the relative frequency of any class we want to look at. Notice that all the z-scores on the table are positive. This is because the area will be the same between a given z-score and the mean, no matter which side of the mean the z-score is on. That is the result of the curve's symmetry. So we look up negative z-scores by ignoring the sign and just looking up the number. Let's do an easy example first.

Example 7:   Suppose we want to know what the relative frequency is for the class of scores between − .42 and the mean. We look up .4 on the table (since it only shows z-scores to one decimal place we round off .42 to .4) and see that the area corresponding to .4 is .1554. We can then round off this number to 16% and say that the relative frequency of the class of scores between .4 and 0 standard deviations is 16%. That is, about 16% of all the scores fall between the mean and .4 standard deviations below (or above) the mean.

Figure 9

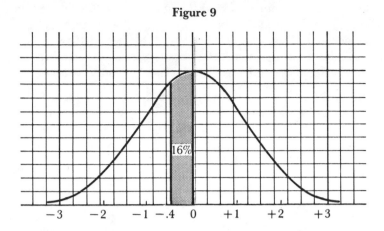

Now we are ready to tackle the more difficult problem of finding the relative frequency of a class that does not have the mean as one of its boundary points. In Example 8 we return to the distribution described in Example 2 and the question we asked there: how do we find the relative frequency of the class of scores between 44% and 73%?

Example 8:   We want to find the relative frequency of the class of scores between 44% and 73% in a distribution with a mean of 66% and a standard deviation of 11%.

First we must find the z-scores for 44% and 70%.

$$z = \frac{44\% - 66\%}{11\%} = -2.0 \qquad z = \frac{73\% - 66\%}{11\%} = .6$$

Next we look up the two z-scores on the table. We find that the area corresponding to −2.0 is .4772 and that corresponding to .6 is .2257. We round off those figures to .48 and .23 and then add. The graph below should make it obvious why we add these two areas. We find that the area of the graph between the score 44% and the score 73% is

$$.48 + .23 = .71 \text{ or } 71\%$$

This tells us that 71% of the total area under the curve falls between the two scores and that therefore the relative frequency of the class of scores between 44% and 73% is .71 or 71%.

Figure 10

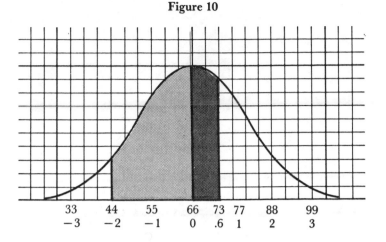

33	44	55	66	73	77	88	99
−3	−2	−1	0	.6	1	2	3

Exercise: What is the relative frequency of scores between 58% and 68%?

In Example 8, the boundary points were on opposite sides of the mean. We found the areas on each side and added them. Suppose we were interested in a class having both boundary points on the same side of the mean. How would we find the area then?

Example 9: The number of detectable defects per new car coming off a particular assembly line turned out to have a mean of 39 with a standard deviation of 5. What percent of the cars had between 40 and 50 defects?

First we find the z-scores corresponding to 40 and 50.

$$z = \frac{40 - 39}{5} = .2 \qquad z = \frac{50 - 39}{5} = 2.2$$

Next we look up the z-scores on the table and find that .2 corresponds to an area of .0793 or .08, and 2.2 to .4861 or .49. Now look at the graph below. Do you see why we **subtract** to find the area between the z-scores of .2 and 2.2?

$$.49 - .08 = .41$$

The relative frequency of the class 40 − 50 is .41 or we can say that about 41% of all cars coming off this assembly line have between 40 and 50 defects.

Figure 11

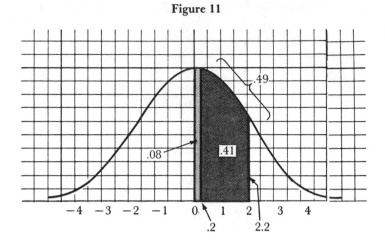

As you can see from the previous two examples, it is possible to solve area or relative frequency problems by applying the rule, "When the boundary points are on the same side of the mean, subtract; when they are on opposite sides, add the areas found from the table." However, it is better to draw a rough graph and see what you are doing than to blindly follow a rule. One final example of the kind of problems we solve with z-scores:

Example 10: Using the same distribution as in Example 9, suppose we just want to know what percent of all cars are likely to have more than 50 defects and what percent are likely to have fewer.

We already know that the z-score corresponding to 50 is 2.2 and the area is about .49. The percent of all cars with fewer than 50 defects is

$$.50 + .49 = .99 \text{ or } 99\%$$

and the number of cars with more defects than 50 is

$$.50 - .49 = .01 \text{ or } 1\%$$

Exercise: Looking at the same distribution as in Example 9, find what percent of the cars had between 35 and 45 defects. Fewer than 35 defects? More than 45?

PART B: Uses and Misuses of Z-Scores

We know how to use a z-score to find the relative frequency of elements of normal distributions whose means and standard deviations are known. We've seen hypothetical examples — using test scores and production defects — of how these methods might be applied. But remember that these methods assume we are dealing with a normal distribution, and a normal distribution is only a mathematical model of an ideal distribution. This presents us with some problems in applying our methods to the real world.

By way of comparison, take our earlier analogy of a circle. Mathematics derived from ideal circles gives us a formula for calculating the area of a circle. But which circular-type objects can we really apply it to? Suppose we want to know the area of a pancake; maybe some efficiency expert has decided to pay short-order cooks according to the number of square inches of pancakes they produce. If a pancake is made by dropping batter into the middle of a smooth, evenly heated griddle, the pancake will approximate a perfect circle; it will be more or less round, and the formula will work fairly well. But if the batter is dropped near the edge of the griddle so it can't spread in all directions, the pancake won't be round and the formula won't work.

Now, what conditions are likely to give us actual distributions that most resemble our ideal model? First, we must have a very large number of cases. Second, we must be measuring something which varies by chance rather than design or principle. For example, height and weight are characteristics of people or animals which tend to vary randomly, and a large enough group of either is likely to fit a normal distribution pattern. Alternatively, as we have pointed out before, income does not vary randomly (it is not pure chance that determines what people earn) so that an income distribution is very unlikely to fit a normal distribution.

Use of Z-Scores

Scores from tests given to very large numbers of subjects tend to form normal distributions. This does not necessarily mean that the ability being tested is randomly distributed in the population, since test scores are affected by a large number of other factors that can vary randomly. By the time ability, interest, motivation, previous exposure, health, energy, distractions, etc. have all contributed to the result, it seems reasonable to assume that there is enough random variability to give something approaching a normal distribution. The large number of people all over the country subjected to these tests makes the use of z-scores feasible. But what are the advantages? In Chapter 2, we talked about two things we wanted scores to tell us — the level of performance with respect to the material and with respect to others in the group. To a certain extent, z-scores give us an indication of both. If we know the mean and the standard deviation (as we usually do in the case of standardized tests) then the z-score contains much information since it is related to both the percent score that tells us how much of the test was completed correctly and the percentile rank which shows how the individual scored with respect to the rest of the group.

Example 11: One of the college entrance tests given on a nationwide basis used z-scores for many years. Specifically, the score a student received was calculated using the formula:

$$S = 100z + 500.$$

To interpret a score of, say, 400, one could then substitute 400 for S in the formula and solve for z:

$$400 = 100z + 500$$
$$-100 = 100z$$
$$-1 = z$$

Thus a score of 400 is one standard deviation to the left of the mean. A score of 700 is 2 standard deviations above the mean.

Since the z-score is figured from either the raw score or the percent score, we can recover those scores when we know the z-score and the mean and standard deviation.

Example 12: Suppose a student's z-score on a nationally standardized test with a mean of 50% and a standard deviation of 10% was 2.3. What was the percent score?

We use the formula from Part A for z-scores but instead of solving for z (which we already know) we solve for x, which we want to know:

$$2.3 = \frac{x - 50\%}{10\%}; 2.3 = \frac{x - .5}{.1}; .23 = x - .5; .73 = x$$

The percent score corresponding to a z-score of 2.3 was 73%.

The z-score is closely related to the percentile rank as well as to the percent score. Remember that the percentile rank is a number which tells us what percent of all scores fall below the score we are considering. In Example 10 of Part A we saw how we could use z-scores and the area table to find the answer to that question. Here is another example.

Example 13: A test given to a large group of students has a mean of 52.5 and a standard deviation of 15. One student scores 84. What is his z-score? What is the percentile rank of his score?

The z-score is:

$$z = \frac{84 - 52.5}{15} = 2.1$$

To find the percentile rank, we first look up the z-score on the table to find the area between it and the mean. The table tells us that area is about 48%. To find the percentile rank, we add the 48% which represents all the scores between 84 and the mean, to 50% which represents all the scores below the mean.

$$48\% + 50\% = 98\%$$
A score of 84 on this test is in the 98th percentile.

Figure 12

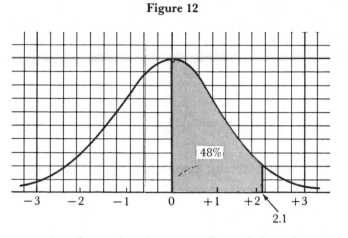

The procedure for getting the percentile rank from the standard or z-score can be summed up as follows: Look up the z-score on the table. If the z-score is positive, add the number under "area" to 50%. If the z-score is negative, subtract the number under area from 50%. In either case, the result is the percentile rank in a normal distribution.

It is important to realize that if the distribution in question is not a normal distribution, this procedure will give completely inaccurate results. Much confusion and distortion results from a tendency to treat every distribution as though it were normal.

Misuses of Z-Scores

A typical example of falsely treating every distribution as though it were normal is the practice, fortunately not as common now as it once was, of "grading on the curve." Teachers who use this method make two assumptions: first that the students in their classes will perform in such a way as to yield test results which approximate a normal distribution, and secondly that the proper way to translate from that normal distribution to actual grades is to assign C's to those with a z-score of +1 or −1, B's and D's to those with z-scores of +2 or −2, and A's and F's to those with z-scores of +3 or −3 or beyond. The first assumption fails primarily on the grounds that no single class is likely to be large enough to provide even an approximately normal distribution. Even if the various factors that go into individual performance create random variation, this is likely to happen only when there are thousands of individuals in question. The second false assumption is that the mean automatically represents C work. There are certain variables affecting all the students equally, such as the amount of information presented to them, the method of presentation, the inherent interest of the subject matter, etc. Thus it seems quite reasonable that one class in psychology at a given university might as a whole learn twice as much as another class at the same university. Why should a student with a score near the mean in the first class receive the same grade as a student scoring near the mean in the second class? Grading on the curve almost always represents a misuse of z-scores and an injustice to students.

The Importance of the Normal Distribution

In this chapter we have looked mostly at one small area in which normal distributions and z-scores are used — the area of test scores. We have not yet touched on the real value and use of these mathematical tools. Now it is time to turn from descriptive to inferential statistics.

In inferential statistics we are not interested in the particular distribution of scores or measurements that has been collected for its own sake. Rather the importance lies in what those measurements reveal about a much larger group than can be dealt with directly. And the normal distribution is the main link in the logical sequence that takes a scientist from the small number of cases he has actually investigated — his sample — to the conclusion he draws about the phenomenon he is interested in. The logical argument he uses is not a simple one, and in order to fully appreciate it, we must look at two intermediate steps first: sampling and probability. In the next chapter we will learn something about how a small group can be chosen to best represent a much larger group. In Chapter 10 we will look at the role of probability in inferential statistics. Then we can put all the pieces together and understand how inferential statistics — the mathematics used by scientists to draw wide-ranging conclusions from small amounts of data — all hinges on the normal distribution.

SUMMARY

PART A

A **normal distribution** is a mathematical model of an ideal distribution which is unlikely to be found in the real world but which is approximated by a variety of real distributions. It has the following characteristics:

1. Its graph has a bell shape.
2. It is symmetric so that the mean, median and mode all fall in the same place at the highest point of the curve.
3. The areas under the curve are always distributed in the same way with about 68% of the area falling within one, 95% within two, and 99.7% within three standard deviations of the mean.

The normal distribution is completely described by its mean and its standard deviation.

A **binomial distribution** is used when considering an experiment or observation with only two mutually exclusive outcomes. The formulas for finding a binomial distribution are:

$\bar{x} = np$ where $\bar{x}$ is the mean, n the number of items in the sample and p the probability of the desired outcome.

$s = \sqrt{np(1-p)}$

The **standard normal distribution** is the normal distribution we get when we use 0 for the mean and 1 for the size of one standard deviation. Any normal distribution can be translated into standard form by the use of z-scores.

A **z-score** is a number corresponding to a data point or a raw score which represents the number of standard deviations that the point is from the mean. The formula for finding a z-score is:

$z = \dfrac{x - \bar{x}}{s}$ where x is the value of the data point, $\bar{x}$ the mean and s the standard deviation.

By using the z-score and the area table we can find the relative frequency of any class in a normal distribution. The area table gives the part of the total area which lies between the score and the mean. The area between any two z-scores can be found by adding or subtracting the areas found on the table.

PART B

Actual distributions are likely to approximate normal distributions only if they are very large and if they measure something which is randomly distributed — that is, governed by chance.

An appropriate use of z-scores is in the area of standardized tests which are given to large numbers of people in a variety of places. Z-scores are useful because they are closely related to both percent scores and percentile ranks.

An example of an inappropriate use of z-scores is in determining classroom grades.

Normal distributions have an importance beyond approximating distributions of standardized tests. Their central role in inferential statistics will be taken up in later chapters.

EXERCISES

PART A

1. *Using graph paper, draw a freehand curve for a normal distribution with a mean of 10 and a standard deviation of 2. Count the squares of

graph paper to see how well you did. Remember that about 34% of the squares should be between the mean and one standard deviation to the left, about 13.5% between one and two standard deviations and so on. The other side should match. If your first attempt was too far off, see if you can make adjustments to improve on it.

2. Take ten coins and toss them as often as you have the patience to do it. After each toss count the number of heads you see. Make a frequency histogram of your results. How close does it come to a normal distribution?

3. *Find the z-score corresponding to a raw score of 72 on a test with a mean of 60 and a standard deviation of 8. Find the z-score corresponding to a raw score of 45 on the same test.

4. What percent of all students in Exercise 3 scored between 45 and 72 on the above test? What percent scored between 72 and 80? What percent scored above 80? Below 72? Below 45?

5. Suppose 10,000 students took the test in Exercise 3. How many of them scored between 45 and 80?

6. Suppose that 2.5% of the students who took the test in Exercise 3 failed it. What was the lowest passing score?

7. *A one-pound box of a particular brand of butter actually contains .98 pounds of butter on the average with a standard deviation of .02 pounds. What are the chances of buying a box that weighs a pound or more?

8. The size of a ball bearing must vary no more than .5 millimeters from the average. If a factory is producing ball bearings with a mean diameter of 6 millimeters with a standard deviation of .3 millimeters, what percent of their ball bearings will have to be rejected?

9. *We are interested in selecting a sample of 50 people. If we assume that the probability of any given member of the sample being a man is 50%, what is the probability that our sample will include 15 or fewer men?

PART B

1. *Graph the data from Chapter 6, Exercise 4 as a frequency histogram, using one standard deviation for the class size. How well does the data fit the criteria of a normal distribution? Be sure to check all three criteria.

2. Suppose the test scores you just graphed were assigned grades on the curve. Which percent scores would get A's, B's, C's, D's and F's? What would be some of the difficulties or injustices involved in using this system?

3. Below are descriptions of a number of different distributions. Which are most likely to approximate normal distributions? Why? How would

you expect those that are not like normal distributions to look?

a. *The weekly grocery bill of a family during a year.
b. The age at which each of 1000 children took their first step.
c. *The birth weights of all calves from a large Texas cattle ranch over a ten year period.
d. The age at death for each inhabitant of New York who died during 1977.
e. *The daily Dow-Jones Industrial Average during the last month.
f. The reading scores of all Seattle 5th graders this year.
g. The incomes of 2 million people selected at random.
h. The sum of the digits on the first 20 license plates seen.
i. The number of times "6" appears in 20 rolls of one die.
j. The number of times "6" appears in 20 rolls of one die if the 20 rolls are repeated 500 times.

Chapter 9

SAMPLING AND OPINION POLLS

PART A: Sampling Techniques

One thing almost all research has in common is that it deals with samples drawn from populations rather than with whole populations. Remember that in statistics the term **population** does not necessarily mean a group of people; it is more likely to mean any set of measurements that have been made or that could be made. A population might be the height of every adult in the United States, the time it would take any laboratory rat to run a particular maze, the number of bushels of wheat grown on any acre of ground, etc. Usually it is impossible to look at every single member of a population. One exception is the U.S. census, which actually attempts to count each individual man, woman and child. But even the census selects a smaller group to answer the longer and more detailed questionnaire that provides much of the data the government publishes. This smaller group is called a sample. A **sample** is any group selected from a population for the purpose of a particular study. Since the use of a sample is just an expedient — it is the whole population we are really interested in — the success of any study depends to a large extent on how well the sample that is chosen represents the whole population.

We are mainly interested in how statistics is applied to the social rather than the physical sciences, so we will concern ourselves here with the problems encountered by social scientists in their efforts to get good samples. And getting a good sample is a major problem for all such researchers. What do we mean by a good sample? Obviously it is one that gives us results as close as possible to those that we would get if we could examine the whole population. But since examining the whole population is just what we can't do, how do we know when we have a good sample?

170

There are no guarantees, but experience as well as certain logical and mathematical considerations do provide us with guidelines. Basically there are two main considerations. First we must get a sample of the right size, and second we must get a random sample. A **random sample** is one chosen in such a way that every member of the population has an equal chance of being chosen. A very simple example will make this concept clear.

Example 1:   Suppose we fill a jar with an unknown but large number of beans, some of which are red and some white. We want to know the ratio of red to white beans.

This problem can be thought of as an analogy to the problem of finding out how a given population will vote on an issue or a candidate. We use beans instead of people because it simplifies the problem and allows us to visualize a procedure. We can imagine taking a sample by simply dipping a hand into the jar and pulling out some beans. We then count the beans of each color and see what the ratio is. Suppose we pull out one white bean and seven red beans. We can then say that the ratio of white to red beans in the sample is 1:7 or we can say that $\frac{1}{8}$ of the beans are white and $\frac{7}{8}$ of the beans are red. Now how well do these sample ratios reflect the actual ratios in the jar? How likely is it that there are really seven red beans for every white bean? It depends on how random our sample was and to a certain extent on how large it was.

To see how random the sample was, we must ask if every bean had an equal chance to be selected. It is easy to see how this might not be the case. If we put all the white beans in the jar first, dump the red beans in after them and then take our sample off the top without mixing the beans, the white beans will have a very small chance of being selected and the red beans a very large one. To get a random sample, we must be careful to mix the beans completely, so that pure chance will determine which end up on top and are selected.

The second consideration is the size of the sample. If we pull out only two or three beans, it is very unlikely that we will get an accurate estimate of the ratio between red and white beans, especially if there are many more of one color than the other. Do you see that the more beans we pull out of the jar, the better our chances are of getting something close to the right ratio? If you have any doubts about that, try it. Put an equal number of beans of two colors in a jar, mix well, and pull out samples of different sizes. Record how many beans of each color there are in each sample you draw. Is it true that the more beans you take from the jar in one sample, the more accurately that sample represents the real ratio? Notice that the question of the size of the sample is quite independent of the question of how random the sample is. If the beans in the jar are not well mixed, increasing the number of beans you look at will not give you a more representative sample. Similarly, if you take a very small sample, regardless of how well mixed your beans are, you will not have enough information to tell you the true ratio of beans in the jar. So how large should a sample be? In a later chapter we will be able to specify more exactly the relationship between sample size and accuracy. For now, we can only say that the size of the sample is one of the considerations we must keep in mind when judging how well it represents the whole population.

The jar of beans is an analogy that gives us an intuitive idea of what we mean by a random sample and a sufficiently large sample. However, we cannot put the people we intend to use for an experiment in a jar and stir them up to make sure we have a random sample. How then are samples really selected? Suppose first that we have a population small enough so that all the members can be listed. We could put slips of paper representing each member of the population in a hat and draw a sample from the hat. The technical difficulties of making sure the slips are really well mixed make this solution impractical. Bingo parlors have machines that are designed to produce randomly selected numbers. We could use such a machine by numbering each member of our population and then getting as many numbers from the bingo machine as we wanted individuals in our sample. The members of the population whose numbers corresponded to the ones produced by the bingo machine would then make up the sample. We don't really need a bingo machine, but we do need a method of getting random numbers. Fortunately these are easily available in tables. A **random number table** is a table of numbers that were generated by a random process, usually by a computer. Such a table is reproduced in the Appendix (p. 305). Here is a list of the steps you would go through to use the random number table to select a sample from a population that is small enough to list.

1. Assign each member of the population a number from 1 to N where N represents the number of people or measurements in the population.
2. Decide how large your sample is going to be. Let n be the size of your sample.
3. Find some random process for selecting a number — any number from the random number table. (You might close your eyes and point, use the number of the day of the week and month of the year to give you a row and column number, or think up some inventive method of your own.)
4. Use only as many digits of the number as there are digits in N. If the number from the table is larger than N, skip it and move on to the next number. The first number you select in this way is the number of the first member of your sample.
5. Move through the number table, either from left to right or top to

bottom, but taking the numbers in some order. Whenever you come to a number that is larger than N or to a number that is the same as one you have already selected, skip it and move on. Continue this process until you have n numbers. If you come to the end of the table before you have n numbers, return to the beginning of the table and keep going. Each number that you pick in this way corresponds to a member of the population who will be in your sample.

Let's do a problem to illustrate how a random table works.

Example 2: Suppose we want to select a sample of 20 students from a freshman class that consists of 350 students.

We will follow the steps outlined above and use the table in the Appendix.

1. Taking an alphabetical list of the freshman class, we assign each student a number between 1 and 350, since $N = 350$.
2. There will be 20 students in our sample, so $n = 20$.
3. Suppose that by some random process we decide to begin with the number in the 7th row and the 3rd column of the random number table. That number is 51135. We use only as many of its five digits as will give us a number that is smaller than or equal to 350, that is, the last three digits. The number we get, then, is 135. The student to whom we assigned the number 135 in Step 1 is the first member of our sample.
4. Now we can move in any direction we like to get our next number. Let's move to the right and proceed row by row. The next number after 135 is 527. This is larger than 350 so we skip it and move on. The next two numbers, 586 and 448 are also too large, but 301 is in our range, so that is the number of the second member of our sample. Proceeding in this way, we select the 20 students who have been assigned the following numbers:
135, 301, 277, 157, 331, 290, 97, 136, 195, 242,
184, 253; 154, 263, 236, 238, 57, 257, 77, 203
Notice that we did not use 236 in column 3 row 18 because we had already picked up that number in column 4 row 17.

To make sure you understand the procedure, try moving down the column from the starting number of 135, instead of to the right. Here are the 20 numbers you should get:
135, 331, 236, 186, 169, 253, 36, 152, 9, 341,
247, 83, 44, 322, 277, 195, 57, 203, 116, 70

Exercise: Instead of starting with the 7th row and 3rd column as we did above, start with the 14th row and 2nd column, and move to the left and list the 20 numbers in the sample.

This method is worth noting because it clarifies what we mean by a random sample. In actual practice, however, it is not often used directly. Let's say we're doing research in reading ability among first graders. We could hardly assign a number to each first grade student in the country and then proceed to study those selected by this method. We might find ourselves testing one child in Tucson, two in San Francisco and maybe a dozen in different schools around New York. It simply wouldn't be practical. What we'd probably end up doing is going to a school where we happened to be friendly with the first grade teacher.

Much actual research uses just such short cuts in selecting a sample, and in some instances that turns out to be quite all right. Whether it does or not depends on the relationship between the sample and what we will call the target population. The **target population** is the largest population to which we are willing to generalize the results of our research. In the case of the reading research example, we might have as our target population all first graders in the US, or all six-year-olds in the world, depending on what we are studying. The sample we picked for our research might differ from the target population in any of a number of ways. The classroom we picked might be in a girls' school, or in an exceptionally rich or poor community. The more relevant differences there are between the sample and target population, the more questionable our results will be. What are relevant differences you may ask?

The ways in which individual members of a sample can differ from each other are called **variables.** When we are dealing with people, we are confronted with hundreds of thousands of variables. Luckily not all the variables are important in any given experiment. One variable among people is difference in hair color. This is obviously irrelevant to our reading experiment. Alternatively, if the sample comes from a specially selected group of exceptionally talented children, their average ability will differ markedly from that of the target population (assuming that the target population is first graders in the US) and that will be an important variable. The question of which variables are important to a particular piece of research is not always so easy to answer. Furthermore, even if we concentrate only on those variables that may have an impact on the research there will usually be more than we can hope to take into account. The best we can do is try to avoid any gross differences between the sample and the target population in those variables that are of obvious importance to the research.

One method of dealing with this problem is to use what is called stratified sampling. **Stratified sampling** is a method that uses random selection techniques while at the same time taking into account some of the essential variables. The researcher must identify these variables in advance. Then the sample is selected so that the individuals in the sample resemble the target population with respect to these variables. For example, if the experimenter feels that the sex of the subjects will effect the result, then the sample will be selected so that half the subjects are male and the other half female, if that is the ratio in the target population. This is done by dividing the population into two **strata,** one containing all males, the other containing all females. Fifty percent of the sample is selected randomly from among the males, and the other half from among the females. If age were an important consideration, the population would be divided into different age groups and then members of the sample would be randomly selected from each age stratum in proportion to the age distribution of the population. When more than one variable is involved, defining the strata becomes more difficult and the number of strata increases rapidly. If we had to consider two sexes and three age groups (say seven, eight and nine year olds) we would need six strata (three seven, eight and nine year old males and three seven, eight and nine year old females).

Example 3: Suppose we want to do an experiment on a new technique to teach reading. Since we are going to compare the results from our sample with those of the rest of the population, we want to make our sample as much like it as possible. We will consider three variables: age, sex and reading ability. Suppose we study 200 seven, eight and nine year olds who, at the end of the first grade, showed average, above average and below average reading skills as measured by a standardized test.

First we must define our strata, then we determine what percent of the population fits each stratum, and finally we select our sample proportionately.

1. We will have $2 \times 3 \times 3 = 18$ strata. Sometimes it helps to make a table like the one below. Each box in the table represents one stratum.

Reading Ability

	Girls			Boys		
	Above Ave.	Ave.	Below Ave.	Above Ave.	Ave.	Below Ave.
Age: 7						
8						
9						

2. Next we must find out what percent of the population fits into each stratum. We consider our population to be all seven, eight and nine year old children in the US. We then need to find out what percent of those children are seven year old girls who are above average in reading in order to fill in the first box, what percent are eight year old girls above average for the second box, and so on until all 18 boxes are filled. The sum of the percents in all the boxes must, of course, add up to 100%.

3. Next, we divide the population from which we will select our sample into 18 categories and select the appropriate percent of our sample from each category. So, for example, if 5% of the population consists of eight year old boys who read below average, then 5% of our sample, or 10 boys (5% $\times$ 200 = 10) must come from that stratum.

4. We select the number of subjects we want from each stratum by some random technique, like using a random number table. Our final sample of 200 children should then be representative of the population at least in terms of age, sex and reading ability.

The basic principles, then, involve finding a sample that is of reasonable size, that is randomly selected, and that is known to be representative of the target population in the variables which are most likely to have an important effect on the question being investigated. The care with which these criteria are met is of particular importance in sampling for opinion polls. In the next section we will look at some of the difficulties that arise with sampling in general and sampling for opinion polls in particular.

Exercise: Suppose an investigator wants to find out how well a new vaccine protects young children (four and five years old) against a common disease. She will consider three variables: age, health and sex. Define how many strata she will need if she uses two age levels and three health levels.

PART B: Sampling Errors

There are a number of techniques besides those explained in Part A for getting a representative sample. There are a great many more ways for failing to get one. We will look at some examples now.

Wrong Sample Size

Perhaps one of the most frequent and obvious sampling errors is that of basing far-reaching conclusions on ridiculously small samples. One can pick up almost any journal at random and see this happening.

Example 4:   The October 1977 issue of *Psychology Today* reports a study which is supposed to show that men and women respond in basically the same way to erotic material. "According to psychologists William Fisher and Donn Myrne, men and women turn on equally, and to the same themes." This conclusion was based on a sample of 31 men and 32 women.

It is very difficult to state what constitutes a large enough sample. Later on, we will learn how to judge the probable amount of error associated with samples of different sizes. But that will not quite answer the question either, because the amount of error that we can live with depends on a number of different factors. Two factors we must consider are the amount of variation we expect in the characteristic we are measuring and the frequency of its occurrence in the population. The amount of variation is important because the larger the sample, the more sensitive it is to variation. For example, suppose we were trying to establish the normal body temperature of a human being. There is very little difference from one person to another, so a quite small sample would be likely to give us a mean somewhere in the neighborhood of 98.6°. Alternatively, if we were trying to establish the average reading speed of an adult population, we would need a very large sample since reading speed may vary from a hundred to several thousand words per minute. Frequency of occurrence is another important consideration. Suppose you are doing cancer research. If the particular cancer you are looking for occurs at the rate of only two per thousand, you will obviously need a very large sample to detect it.

Is it ever possible to have too large a sample? Some statisticians feel that it is. Suppose we are looking at two groups of students, each of which has been taught to read by a different system. We want to know if one system works better than the other. If the groups are small, then the amount of error to be expected will be large, so only a very large difference between the performance of the groups will count as evidence for the superiority of one method over the other. Now as we increase the size of the groups, the amount of error will decrease. Perhaps there is a point at which a really insignificant difference in performance will count as evidence because small as it is, it is still larger than the expected amount of error. Although this situation is theoretically possible, it is hardly a common problem, since the time and money needed to do any research make it relatively unlikely that too large a sample will be studied.

Self-Selected Samples

There are many ways in which a sample can fail to represent the population being studied. One of the more common is that the sample is self-selected. That is, instead of the researcher selecting her sample in some random way, the subjects of the experiment or opinion poll select themselves. To a certain extent, there is always an aspect of self-selection in any research. The subject has to at least co-operate, and by refusing to do so can select herself out of the sample. In some cases the number of people choosing to cooperate is so high that this factor is insignificant. For example, less than two percent of the population refuses to cooperate with the US census takers. However, questionnaires sent by mail are heavily subject to this bias.

Example 5:   In the 1940s, the Federal Trade Commission banned the use of the following ad: "Do you know that the 1940 National Survey conducted among thousands of dentists revealed the following remarkable fact — Twice as many dentists personally use Ipana Tooth Paste as any other dentifrice preparation. In a recent nationwide survey, more dentists said they recommended Ipana for their patients' daily use than the next two dentifrices combined." What the ad failed to state is that out of the 66,000 dentists questioned, less than 20% replied at all. Only 621 dentists disclosed their personal preference, and only 461 revealed what they customarily recommended to patients. This case is discussed in *A Primer on the Law of Deceptive Practices* by Earl W. Kintner (Macmillan, 1971, p. 153).

In the above example it is not clear that this self-selection procedure would necessarily bias the results. The problem is more likely to be that the resulting sample is much too small to be representative. In other cases it is more obvious that those people who select themselves do so because of their bias. Perhaps the most extreme example of this kind of thing is the following.

Example 6:   It has become more and more customary to depend on refugees fleeing a particular country to provide information about conditions in that country. For example, *The New York Times* of September 20, 1977 reports, "Now that southern Vietnam has become virtually impenetrable by foreigners and only the Hanoi Government's picture of life in the reunited country is presented to the world, the thousands of Vietnamese who have fled in small boats are the principal source of critical first-hand information about their country." It is hard to see why information provided by political refugees from a country would be any less biased in one direction than the government's statements would be in the other. In this particular case dependence on either source is unnecessary since journalists from countries other than the U.S. are permitted in Vietnam and so more objective information is regularly available in the foreign press.

Convenience Samples

Self-selected samples are by no means the only non-random samples. Another common type is the so-called **convenience sample.** This is just a fancy name for the kind of sample we encountered in Part A when we talked about using, as a sample for our reading research, the classroom of a friendly teacher. It is a sample used because one has convenient access to it. The use of convenience samples is bad enough in itself but when they are passed off as random samples they lead to even greater falsification.

Example 7: In *Flaws and Fallacies in Statistical Thinking,* Stephen K. Campbell gives the following example. "In this (fairly recent television) commercial, a man runs out into the street and asks each of three people in cars — apparently waiting for the light to change — whether he or she uses a certain product. Two answer 'yes' and one answers 'no.' The viewer, unless he is on his guard, not only tucks away the impression that two out of three people use this product but also that the sample is random. The interviewer picks out three cars quite arbitrarily, hence, the seeming similarity to random sampling. But this is still a convenience sample and not a random one. A truly random sample would give the stay-at-homes and people in other parts of the city a chance to be included as well as drivers in that particular neighborhood." (p. 144)

Another way to look at this problem is to consider what is called the parent population of the sample.

The Parent Population

We have talked about the target population as the population to which we want to generalize the results of a study and the sample as the people we actually look at, but there is an intermediate group called the **parent population.** This is the actual group we select our sample from. As we pointed out earlier, it is hardly ever practical to take the whole population and select randomly from it. If we are doing research on reading, we will choose people who live in our city to work with, not people scattered all over the U.S. The group that we choose our sample from is the parent population. In Example 7 above, the parent population was the people in cars who stopped at the light during the time the interviewer was there. It should be obvious that the more nearly the parent population represents the target population, the more accurate our results will be. Choosing the wrong parent population can lead to disaster.

Example 8: Probably the most famous example of an inappropriate parent population is the 1936 *Literary Digest* poll which predicted that Alf Landon would beat Roosevelt. Roosevelt got 60% of the vote and the *Digest* went out of business. As Michael Wheeler explains in *Lies, Damn Lies and Statistics,* part of the problem was of the type encountered in Example 5: only 15% of those polled responded. The main difficulty, however, was with the parent population: "The mailing lists the editors used were from directories of automobile owners and telephone subscribers. Although the same sorts of lists had provided reasonably accurate results in the past, they were clearly weighted in favor of the Republicans in 1936. People prosperous enough to own cars have always tended to be somewhat more Republican than those who do not, and this was particularly true in the heart of the Depression." (p.69)

It is interesting to note that telephone lists are still used in opinion polls. The theory is that now the telephone is so common in all households that using directories cannot induce a bias. However, one may wonder whether the current trend toward unlisted phone numbers might not induce a new and less well understood bias. We don't know enough about people who don't want their numbers listed or their motives to make a judgment, but it seems to be an area the pollsters should investigate. In some cases it might be quite appropriate to conduct a mail or phone poll, while in others it leads to inaccurate results. It all depends on the question that is being studied.

An inappropriate method of access (mail, phone, etc.) is one way of ending up with a biased parent population. Geographic considerations may be another.

Example 9:   In "For Whom the Polls Toll" (*The Real Paper,* September 11, 1974) Robert F. Duboff states that: "People who live in the following areas are never interviewed: 1) Alaska and Hawaii; 2) College dormitories, old age homes, hotels, etc.; 3) Central cities and other "dangerous" areas; 4) Places distant from where any interviewer lives."

Conservative Bias

The above considerations, when added together lead to a kind of bias that seems almost unavoidable. Darrell Huff describes it as follows in *How to Lie With Statistics:*

You have pretty fair evidence to go on if you suspect that polls in general are biased in one specific direction, the direction of the *Literary Digest* error. This bias is toward the person with more money, more education, more information and alertness, better appearance, more conventional behavior, and more settled habits than the average the population he is chosen to represent.

. . . it is not necessary that a poll be rigged — that is, that the results be deliberately twisted in order to create a false impression. The tendency of the sample to be biased in this consistent direction can rig it automatically (p.26).

How does this happen? There are several clues. Besides the factors already discussed, the following example gives us another insight.

Example 10: In an article entitled "One hour of questions shapes years of policy" in the June 1, 1976 issue of the *Boston Globe,* the following sampling procedure is described for a detailed survey carried out by the Census Bureau for the U.S. Department of Health, Education, and Welfare. "The respondents are chosen at random by computer from what Johnson calls a 'giant' mailing list of addresses. Sometimes when the interviewer gets there the respondents have left, or the house has been demolished, he said, and infrequently the respondent is reluctant to answer questions. 'But we have about 90 percent cooperation.' "

It is not hard to figure out what segment of the population the missing ten percent is most likely to represent. Working from mailing lists which were not compiled yesterday introduces a bias in favor of stability, if nothing else.

Furthermore, we might stop for a minute to consider who has the time or the inclination to fill in long questionnaires received in the mail. Who is likely to discuss their political opinions on the telephone after a hard day's work, or in the middle of a busy afternoon? Who is likely to stop on the street to give information to an opinion taker? On the whole it is clear that the poorer segments of the population are often less accessible, sometimes less interested and frequently less available for polling purposes. The bias is naturally toward those who have a better life.

Opinion Polls

Since the kind of sampling we've been talking about here is of primary importance in opinion polling, we will take a moment here to point out some of the difficulties with polls that go beyond finding a representative sample. These are problems that crop up not only on opinion polls but on any kind of research which depends on people being **questioned.** The problems involve the framing of the questions to be asked, the method of asking the questions, and the nature of the person doing the asking.

Framing the Question. It is not necessary to be as obvious as "When did you stop beating your wife?" in order to prejudice the answer to a question.

Example 11: The following quote is from the *Real Paper,* September 11, 1974:

Essentially, the questions are written in good faith by the pollster, and based on her/his own expertise. However, the wording makes a tremendous difference. For instance, Gallup did a poll on abortions. He asked: "The U.S. Supreme Court has ruled that a woman may go to a doctor to end pregnancy at any time during the first three months of pregnancy. Do you favor or oppose this ruling?" Forty-seven percent approved; 44 percent opposed. A pro-abortion group subsequently asked the same question but deleted "to end pregnancy" and substituted "for an abortion." Their version showed 41 percent approved and 48 percent opposed, and, in a second survey, 53.7 percent opposed and 42.9 percent approved. And, when they simply asked, "as far as you yourself are concerned would you say you are for or against abortion, or what do you think?" they found 59.4 percent against, 36.2 percent in favor. The point is that the way you ask a question can determine the response you get.

Form of the Answer. The way in which the person being interviewed must answer also has an effect. If the interviewer asks a very open ended question, such as "How do you feel about pollution?" she will get such a variety of answers that they will be impossible to analyze and report. Alternatively, when the person being interviewed is given a choice of answers, the available choices often bias the response. As we mentioned in Chapter 5, Example 6, in one study only 3 out of 40 people gave answers to an open-ended question that could be placed into one of the three categories offered when the same question was asked in a multiple choice format.

Another example is a question posed by a Harris Survey interviewer in 1976 about Henry Kissinger. People were asked to agree or disagree with the statement: "Since Henry Kissinger failed to make peace between Egypt and Israel, it looks as though he is losing his touch as a peacemaker." One person complained that she had never thought much of Kissinger's "touch" as a peacemaker in the first place. Answering "disagree" would be taken as an endorsement of Kissinger, since she would be saying, in effect that he still did have the "touch."

On the other hand, she couldn't bring herself to "agree" with the statement, because to do so would imply that, until his Middle East fiasco, she had supported him. As she couldn't subscribe to either of the offered alternatives, the interviewer put her down as "not sure," when in fact she had a clear and strong opinion about Kissinger. As this example shows, most complex attitudes are forced into neat little "yes" or "no" boxes, turning up poll results that are, to say the least, misleading.

This does not mean that it is impossible to design a questionnaire that will get at people's opinions, only that it is extremely difficult to do so. When framing questions with a choice of answers, it is important to consider all possible responses, possibly including such choices as "I don't understand the question" "I don't have enough information to answer it" and "I couldn't care less!"

Effect of Interviewer. Not only the way in which a question is worded or the form in which the answer is recorded will affect the response. Who is asking the question turns out to be important as well.

Example 12: In *Statistics: A Guide to the Unknown* (Holden-Day, 1972) the following point is made.

> "It had long been known that enumerators (census takers) can and do influence the answers they obtain — presumably, unconsciously most of the time. But the magnitude of this enumerator effect was not known. Hence a large statistical study was carried out as part of the 1950 census to measure the magnitude of enumerator effect.
>
> . . . Far greater differences between enumerators were found than had been anticipated, not so much on items such as age and place of birth, but on the more difficult items such as occupation, employment status, income, and education. For those items, in fact, a complete census would have as much variability in its results (because of enumerator effects) as would a 25% sample if there were no enumerator effect!" (p. 283)

A vivid example of just how the interviewer can affect the outcome of a poll without any intentional biasing is given in *How to Lie With Statistics* (p.24).

Example 13: During the Second World War, "the National Opinion Research Center sent out two staffs of interviewers to ask three questions of five hundred Negroes in a Southern city. White interviewers made up one staff, Negro the other. One question was 'Would Negroes be treated better or worse here if the Japanese conquered the U.S.A.?' Negro interviewers reported that nine percent of those they asked said better. White interviewers found only two percent of such responses. And while Negro interviewers found only twenty-five percent who thought Negroes would be treated worse, white interviewers turned up forty-five per cent.

When 'Nazi' was substituted for 'Japanese' in the question,

the results were similar.

The third question probed attitudes that might be based on feelings revealed by the first two. 'Do you think it is more important to concentrate on beating the Axis, or on making democracy work better here at home?' 'Beat Axis' was the reply of thirty-nine percent, according to the Negro interviewers; of sixty-two per cent, according to the white."

Over the years, polling techniques have improved considerably. Pollsters have learned from their mistakes and computers have been put to good use so that the analysis of election results as they come in on election night often seem almost miraculously accurate with only a tiny fraction of the precincts in. So the techniques are there. Whether they are used or willfully misused is another question.

Polling to Support Positions. So far we have talked about polling errors that may or may not be intentional. There is at least one experienced pollster, Albert Sindlinger, who claims that polls are often consciously rigged.

Example 14: In *Lies, Damn Lies and Statistics,* Sindlinger is quoted as saying:

> People hired private pollsters to prove a point — no one's going to pay someone to prove they're not number one.' According to him, many businesses determine the course of corporate policy and then commission a poll to justify their decision. If their action does not pay off, they can always make the excuse that it was based not on their own judgment but on solid research. (p. 237)

In this case, the only loser is the business which fails or loses profits and the workers it must lay off. A more serious problem might be in the area of political polling. People are becoming more and more suspicious that pre-election polls might actually influence election results. Everyone wants to back a winner. This possibility is examined in an article entitled "Do polls influence voters or candidates?" in the New England section of the *Boston Sunday Globe* of June 20, 1976.

Example 15: Albert Sindlinger, who worked for Gallup in the 1940's, accuses Gallup of rigging his polls in favor of Dewey. "We'd set up the headlines and draft the story and then we would go out and do the surveys to fill in the gap," Sindlinger is quoted as saying. "If the results squared with our story, we'd congratulate ourselves on how smart we were. But if they didn't, then the data would be adjusted, supposedly because there was something wrong with the sample."

Whatever techniques are used, however honestly, we must keep in mind that whenever the result of an experiment is based on a sample, the result is *true* only of that sample. As far as the rest of the population is concerned, the result is, at best, an estimate. In the next chapter we will look at the mathematics of gauging how good an estimate it is likely to be.

SUMMARY

PART A

A **population** in statistics is any set of measurements that have been or could be made. A **sample** is a group selected from a population for the purpose of a particular study since the population is usually too large to be examined completely. The **target population** is the group to which the results of the study of a sample are to be generalized. In order for the sample to represent the target population accurately, the sample must be large enough and it must be randomly chosen.

A **random sample** is one chosen in such a way that every member of the population has an equal chance of being chosen. One way to do this is to assign each member of the population a number and then use a **random number table** to select the sample.

Stratified sampling takes into account some of the relevant variables in the population. The population is divided into strata, or subgroups, each of which displays a different set of characteristics. The sample is then chosen by using random methods to select the proper number of subjects from each stratum.

PART B

One of the most obvious kinds of sampling errors is to use **too small** a sample. It may also be possible to use **too large** a sample. **Self-selected samples** or samples consisting of volunteers are likely to present a false picture, as are **convenience samples,** those that are chosen in a non-random fashion because they are convenient. The group from which a sample is chosen is called the **parent population,** and it is the relationship between this population and the target population which determines the value of a randomly selected sample.

All opinion samples tend to have a **built in bias** because there are certain segments of the population which rarely participate in polls or surveys.

Opinion polls can go wrong on other than sampling inadequacies. Answers are often influenced by **how the question is framed, who is asking it,** whether it is solicited by means of **written questionnaire, telephone or personal interview.** It is also sometimes possible that polls are consciously and purposely **rigged.**

EXERCISES

PART A

1. *Suppose you decide to do research on how college students feel about nuclear energy. You decide to select a sample of 15 students from the freshman class of your college. There are 550 students in the class. Using the technique of selection by use of the random number table, answer the following questions.
   a. What is the value of N in this example?
   b. What is the value of n?
   c. How did you select the first number on the random number table?
   d. What are the 15 numbers you found?
   e. How do you determine who the 15 students in your sample will be?

2. After doing the experiment in Exercise 1, you decide to follow up with a larger study. You decide to use 50 subjects and to stratify your sample. You decide the variables you want to consider are sex, political party affiliation (Democrat, Republican and Independent), and whether or not the person has ever voted.
   a. How many strata will you need?
   b. How do you decide what percent of the population or sample fits into each stratum?
   c. How many of your subjects will be women?
   d. If 50% of the population are Democrats and 25% have never voted, how many of your subjects will be Democratic women who have never voted?
   e. Make up your own data to fill in a table with percents in each stratum.

3. *If you were doing a survey on housing needs in your community, what are the three variables you would be most likely to consider important? What are three variables that you would consider irrelevant?

PART B

1. In Exercise 1 above, what is the target population, and what is the parent population?

2. *What criticisms can you make of the poll in Example 1?

3. What criticisms can you make of the following procedures?
   a. From the *Boston Sunday Globe* of June 20, 1976: ". . . John Becker, who does the polling for Massachusetts GOP Sen. Edward W. Brooke and other state Republicans, prefers the telephone. . . . The use of the telephone is a good argument for saving money. It is fast and can produce numbers quickly."
   b. *In *The Golden Fleece* Joseph J. Seldin says, "A *Tide* survey, for instance, found three out of four ad executives confessing that they did

conduct informal polls among their wives, friends, neighbors, or whoever else was handy," and that this influenced policy decisions.

c. The *Reader's Digest* of February 1969 reported the following piece of research. "For example, a recent study of 370 asbestos workers showed not one case of lung cancer among the non-smokers, whereas lung cancer among the smokers was much more prevalent than would be expected in a normal male smoking population."

d. A poll quoted in the *Boston Globe* of Dec. 2, 1973 asked the following question. "Which one of the following do you think is most responsible for our present fuel and energy situation:
   — the public itself, by using and demanding too much oil and electricity;
   — our support for Israel against the Arab countries;
   — the Nixon Administration, by failing to recognize and begin taking steps to prevent the shortage early enough;
   — the major oil companies by refining too little and exporting too much oil; or
   — Congress, by waiting too long to give the President the necessary authority to deal with the situation?"

4. What criticisms can you make of using the following questions for polling purposes?
   a. *What steps do you think Congress should take to reduce the amount of taxpayers' money being spent on welfare payments?
   b. Do you agree that children have a right to be protected from pornography?
   c. *Do you agree that most politicians are dishonest?

5. Suppose you are taking a poll to determine public opinion on the Equal Rights Amendment. Briefly outline possible sampling procedures, data gathering techniques, types of questions and answers and other factors that you would use under each of the following three conditions:
   a. You want the poll to show a positive attitude toward the ERA.
   b. You want the poll to show a negative attitude toward the ERA.
   c. You want to get the most accurate possible picture of public opinion.

Chapter 10

PROBABILITY AND SAMPLING DISTRIBUTIONS

In the last chapter we saw that most research must be done on a sample which represents a much larger population. We talked about some of the precautions which can be taken to make that sample resemble, as much as possible, the target population or the group which we are studying. Now we must deal with a factor which affects how similar our sample actually is to the population, a factor which we cannot control. That factor is chance.

Consider Example 1 in Chapter 9. We had a jar containing red and white beans and we wanted to know the ratio of red to white. We mixed up the beans thoroughly and pulled out a handful of beans for our sample. By mixing the beans we had taken every possible precaution to see that the procedure was random — that each bean had as good a chance of being selected as any other. Yet our sample could have included only red beans. Common sense and experience tells us that it is perfectly possible that by chance, we can miss including any white beans in our sample, and so by chance we can get a misleading result. Whenever we use a sample to represent a whole population, we are taking that risk. For example, suppose we are interested in finding out what the average height of an American is. We draw a sample but it turns out by chance that our sample includes a large number of basketball players. Despite the fact that our procedure for selecting the sample was random, we still came up with a sample which was unrepresentative of the population. As we can see from this example, we might make a small error or a very large one, entirely by chance. In the next three chapters we will look at ways of assessing that risk. We will use mathematics to help us decide how likely our sample is to produce an incorrect estimate of the measure we are interested in, and how large that error is likely to be.

PART A: Examining Chance

In order to understand the role that chance plays in statistical research we must learn a little about probability. **Probability** is the branch of mathematics which studies chance. We use the word probability in everyday speech in a variety of ways, some more technical than others. We might say, "There's a strong probability that the Phillies will win the World Series." The weather forecaster might say, "The probability of precipitation is 80% tomorrow." The card dealer might say, "The probability that the next card I turn over will be a spade is 25%." All of these statements can be interpreted as measures of the strength of belief in the outcome of an event. Many different theories have grown out of attempts to measure that belief. In the following discussion, we will talk about only one of those theories: the relative frequency approach to probability. Before we can understand what that means, however, we will have to define a few other terms.

We will use the word "experiment" in this chapter to mean a simple action, such as the toss of a penny or the drawing of a card, the outcome of which we are interested in calculating probabilities for. It is an action that can be repeated any number of times and that can have different outcomes. Each possible outcome is called an **event**. An event may also be thought of as the result of an observation. A **simple event** is one which meets only one condition; a **compound event** must meet two or more conditions. So if we perform the experiment of drawing a card from a deck, we call drawing a spade a simple event, while drawing the queen of spades is a compound event, since it must meet the condition that the card be a queen as well as that it be a spade. Our task will be to calculate the probability that a certain experiment will result in a certain event.

Theoretical and Empirical Probability

The two most common kinds of probability are the theoretical and the empirical. The difference is that in the case of a theoretical probability we already know in advance what the different possibilities in a particular situation are. The simplest example is that of tossing a fair coin. We know that there are only two possibilities. A tossed coin can either land "heads" or "tails." We don't need to actually toss the coin to reach the conclusion that the probability is 50% for the coin landing "heads" and 50% for "tails." The probability is theoretically clear. In some situations, however, we have no way of knowing in advance what the possible outcomes are, or we do not want to prejudge them. For example, we may have a coin that we suspect is weighted so that it comes down more frequently "heads" than "tails." Then we must resort to the empirical notion of probability. The word **empirical** means depending on experience or observation rather than on theory. In order to find the empirical probability of our doubtful coin landing "heads" we must actually observe what it does, so we toss the coin a large number of times and record the number of "heads" we observe. The empirical probability is equal to the relative frequency of "heads" — that is, the number of times the coin landed "heads" divided by the number of times it was tossed. Since the result of numerous repetitions of this experiment will come out a little differently each time we try it, any actual result is only an approximation of the empirical probability. The **empirical probability** of an event is the relative frequency with which that event would occur if the experiment were repeated infinitely many times. The more often an experiment is actually tried, the closer we are likely to come to that ideal empirical probability.

Whether we are using theoretical or empirical data for calculating probabilities, we can still use the relative frequency method. For theoretical probabilities, we divide the number of possible ways in which the desired event (the one we are looking for) can occur, by the total number of possible events. For empirical probabilities, we divide the number of occurrences of the desired event by the number of times the experiment was performed. We can use a single formula to express the probability of a desired event occurring:

$$p(e) = \frac{n(e)}{n(t)}$$ where $p(e)$ is the probability of the desired event, $n(e)$ is the number of times it can or does occur, and $n(t)$ is the total number of possible or actual experiments.

Notice that $n(e)$ cannot possibly be larger than $n(t)$; therefore $p(e)$ will always turn out to be no more than one. Also, $n(e)$ can never be less than 0, so $p(e)$ will always be at least 0. Therefore a probability is always a number between one and zero. The following two examples should make the use of this formula clear.

Example 1: Suppose we want to know the theoretical probability of drawing a spade from a pack of cards. $p(e)$ stands for the probability of drawing a spade; $n(e)$ stands for the number of times the desired outcome occurs or the number of different ways it can occur. Since there are 13 spades in a deck of cards, $n(e) = 13$. Since $n(t)$ is the total number of different possible outcomes, and there are 52 cards in the deck, $n(t)$ is 52. Therefore we have:

$$p(e) = \frac{13}{52} = \frac{1}{4} = .25$$

Example 2: Suppose we have a jar containing white and red beans but we do not know how many of each color there are. We want to know the probability of drawing a white bean from the jar. We will decide the probability empirically by removing and counting a large number of the beans. The larger the number we count, the closer our estimated empirical probability will be to the actual probability. Suppose we decide to count 100 beans and find 38 of them to be white. We can then use our formula to find the probability that a single bean drawn from the jar will be white.

$$p(e) = \frac{38}{100} = .38$$

Exercise: Suppose an urn contains 25 black balls and 60 white balls. If we select one ball, what is the probability of its being black?

So far we have looked at simple events only, such as drawing one white bean from a jar or having one penny come up "heads." Now we want to consider **compound events,** that is, events made up of two or more simple events. We might ask questions like: What is the probability of a penny coming up heads three times in a row? What is the probability of drawing a card which is both a face card and a heart? What is the probability of meeting a black congresswoman? When we are concerned with these more complicated probabilities, we have to be aware of the relationship between the different simple events that make up the event we are interested in. Either the simple events are **independent,** meaning that the probability of one event happening has no influence whatsoever on the others, or the simple events are **non-independent,** meaning that the occurrence of one of the events may increase or decrease the probability of the others.

When we calculate the probability of a compound event, we must know whether the simple events are independent or not because the calculations are quite different. Unless the two events are independent, the calculations are generally too complex to discuss here. If the events are independent, then the probability of the compound event is quite simple to calculate from the probabilities of the simple events that make it up: the simple probabilities are multiplied together. In the case of a compound event made up of two independent simple events, the formula would be:

$$p(e) = p(e_1)p(e_2) \text{ where } p(e_1) \text{ is the probability of one simple event and } p(e_2) \text{ is the probability of the other.}$$

If the compound event is made up of more than two independent simple events, then all the probabilities of the simple events are multiplied together.

Example 3:   What is the probability of a penny coming up heads twice in a row? Since the result of the first toss in no way influences the result of the second, the two events are independent. As the probability of each simple event is .5, the probability of the compound event is .5 × .5 = .25.

Exercise: What is the probability of a penny coming up heads four times in a row.

If two events are not independent, then the probability of their both occurring cannot be figured by multiplying their individual probabilities. Compare what happens in the two cases below.

Example 4: If we ask what the probability is that someone will draw an ace on each of three consecutive attempts when the card that was drawn is replaced in the deck, then we have three independent events and the probability is $(\frac{4}{52})(\frac{4}{52})(\frac{4}{52})$ = .00046. Alternatively, if the drawn card is *not* replaced, then the three events are not independent and the formula will not work. This is because after the first draw the deck is reduced to 51 cards and if the first draw is successful, then only three rather than four aces remain in the deck, making the chance on the second draw, 3/51 (5.9%). If that draw also is successful the probability of success on the final draw is 2/50 (4%). But if any of the draws are not successful then more aces remain in the deck and the probability on the following draw is different. So how do we know which probabilities to multiply?

Figuring probabilities of non-independent events is beyond the scope of this book. For our purposes it will be sufficient to recognize the difference between the two types of events and to remember that non-independent probabilities must be calculated differently from independent ones.

Whether the probability we use is arrived at theoretically or empirically, whether it is for a simple or compound event, the point of a probability is that it is used in order to predict future events. The theoretical probability that any coin comes up heads 50% of the time allows us to predict that any particular toss has a 50% chance of landing heads up. The empirically determined probability that our weighted coin has an 80% chance of landing heads up allows us to predict that any future toss of that coin has an 80% chance of landing heads up.

Probability Distributions

We use probability in statistics to help us estimate how well a sample represents a population. In order to do that, we need to work with probability distributions rather than individual probabilities. A **probability distribution** is a distribution showing the probabilities associated with each of the possible outcomes of an experiment or observation. Although the term "probability distribution" is new to us, the concept is not. Any relative frequency distribution that presents data for a whole population can be considered a probability distribution.

Example 5:   Suppose we wanted a probability distribution for income in the U.S. in 1975. We would use empirical data, from the *U.S. Statistical Abstract,* and our distribution would look something like this:

Figure 1

Probability Distribution for U.S. Family Income

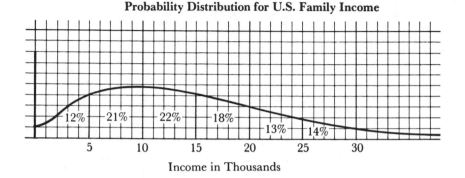

Income in Thousands

Like all empirical probability distributions, this one is simply the smooth curve that best approximates the histogram for the actual population. It can be used to predict the probability that an individual, chosen from the population at random, will fall into any particular class we may be considering.

The above example is an empirical distribution determined by an actual population. What we usually mean when we speak about probability distributions, though, is a theoretical distribution that we can use as a model for a great many different populations. And this, of course, is precisely the type of distribution we encountered in Chapter 8 — the normal distribution. Although there are many different kinds of probability distributions, the most important ones from our point of view are normal distributions. Consider the following example.

Example 6:   Suppose our population consists of all the people taking a standard college entrance examination in one year. The mean for the examination is 500 and the standard deviation is 100. We expect the scores to be normally distributed, so the probability distribution of the scores will look like this:

Figure 2

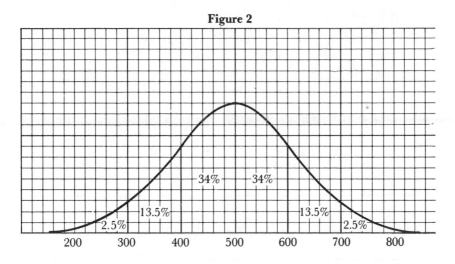

This distribution tells us what the probability is that any particular score will fall within any class we want to look at. The probability of scoring between 500 and 600 is about 34%, the probability of scoring over 700 is 2.5%, and so on. If we now looked at the actual distribution of scores for the year in question, it would be quite a lot like this normal distribution, but of course it would not be identical to it. The probability distribution is a model of the population distribution.

The above example shows a probability distribution for **numerical data** — data we get from making numerical measurements, such as height, age, reaction time, etc. Sometimes, however, we want to deal with data that cannot be expressed numerically. For example, we might be interested in whether individual members of a population are male or female, Democrats, Republicans or Independents. We might want to know whether a card is red or black, or a coin comes up "heads" or "tails." If we have only two mutually exclusive categories to deal with, then the binomial distribution serves as our model probability distribution. We encountered the binomial distribution in Chapter 8. It is another example of a probability distribution.

Now why do we need these theoretical probability distributions? We need them to help us answer the question we began this chapter with: how can we judge whether a sample we have selected actually represents the population from which it was drawn? In order to answer that question we will now look at a particular type of probability distribution called the sampling distribution.

Sampling Distributions

A sampling distribution is a probability distribution which shows the probability of getting a sample having a particular statistic from a given population. Suppose the statistic we're interested in is the mean. When we looked at the distribution of examination scores in Example 6, or of incomes in Example 5, we were concerned with the probability that an individual would fall within a certain class of scores or incomes. Now we're interested in what the mean score or income of a sample of those individuals is likely to be. Instead of looking at the probability that a single randomly selected U.S. family will make $15,000 a year, we want to know what the probability is that a randomly selected group of U.S. families will have a mean income of $15,000.

"Sampling distribution" is a confusing term. Bear in mind that it does not refer to the distribution of income or scores that we will find *within* a sample. It refers to the distribution of mean incomes or scores *among* many samples. It is a description of how much the samples will differ from each other. It allows us to predict the likelihood that a random sample from the whole population will have a certain mean.

It is actually possible to construct sampling distributions for many measures of what a sample may be like, not just for the mean of the sample. But here we will look at only two — the sampling distribution of the mean, and the sampling distribution for binomial probabilities. That will allow us to look at both numerical data and data for either-or categories such as responses to opinion polls.

All this will become clearer when we look at specific examples. First, however, we must take a moment to discuss notation. In constructing formulas to reflect the relationships of samples, populations, and sampling distributions, we must refine our symbols for means and standard deviations to show which mean and which deviation we are talking about. To that end, we will use the following notation, which is standard to most statistics texts.

$\bar{x}$ means *the mean of the sample*
s means *the standard deviation of the sample*
μ means *the mean of the population*
σ means *the standard deviation of the population*
s.e. means *the standard deviation of the sampling distribution*

We have already encountered $\bar{x}$ in Chapter 6 and s and σ in Chapter 7. The Greek letter μ (mū) is the Greek version of m and stands for the mean. The abbreviation *s.e.* stands for standard error, and we shall soon see why that is an appropriate name for the standard deviation of the sampling distribution.

Sampling Distribution of the Mean. We are looking to see how, when we take samples from a given population, the means of those samples are distributed. Let's look again at the population from Example 6.

Example 7:   This population consists of test scores with a mean of 500 and a standard deviation of 100. Now suppose we select many samples, each of 25 scores, from this population and look at the distribution of the means of these samples. That is, we take the mean of each sample of 25 scores, and record those means on the horizontal scale of our graph. If we take enough samples, our distribution will be approximately normal. We can graph it like this, and we will be able to use our knowledge of areas under the curve and z-scores to make calculations about probabilities.

Figure 3

Distribution of Sample Means

N = 25

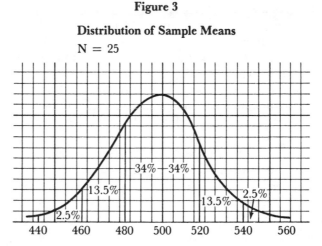

Like any other normal distribution, a sampling distribution can be defined by giving its mean and its standard deviation:

1. The mean of the sampling distribution is always the same as the mean of the population. When the mean of the population is known, we will use μ for the sampling distribution's mean. Where it is being approximated by the mean of a sample, we shall use $\bar{x}$.

2. The standard deviation of the sampling distribution, or the standard error, is given by the formula:

$s.e. = \dfrac{\sigma}{\sqrt{n}}$ where σ is the standard deviation of the population and n is the size of the sample.

Where do these formulas come from, and what do they tell us? Let's consider a concrete example.

Example 8:   Suppose we have a very large apple orchard. We weigh each apple as it is picked and also count the total number of apples. From this data we calculate the average weight of an apple and the standard deviation. We find that $\mu = 6$ oz. and $\sigma = 2$ oz. Then we pack the apples into bags of 25 apples each. If we consider each of these bags a sample, what does the sampling distribution look like?

Mean. It should be fairly obvious why the mean of the sampling distribution is the same as the mean of the population. The weights of the apples haven't changed, we've just grouped them differently. In the case of the population mean, we've added up the weights of all the apples and divided by the number of apples there are. To find the mean of the sampling distribution, we would add up the weights of the apples in each individual bag and then divide by 25, the number of apples in the bag. After we have a mean for each bag, we would add up all those means and divide by the number of bags, thus finding a mean of all those means. That gives us the mean of our sampling distribution. But in either case, we are still finding the typical weight per apple, so our answers should be very close, if not identical. When we are actually talking about a theoretical sampling distribution, it is more as though we made up one bag of 25 apples, took the mean, emptied it back into the pile of apples, made up the next bag, and repeated this process infinitely many times. That way all our samples would be truly random because each apple would have an equal chance to be chosen in each sample. Under those conditions the mean of the sampling distribution would be exactly the same as the mean of the population. In this case we see then that $\mu = 6$ oz.

Standard Deviation. The standard deviation of the sampling distribution tells how much the means of the individual samples will spread out around the mean of the population. This spread is influenced by the spread (standard deviation) of the population. The more variation there is in the size of apples in the population, the more variation there will be between one sample and another. But the standard deviation of the population isn't the whole story. The size of the sample is important, too. If we take very small samples, there will be little chance for the largest and smallest apples to balance out within any one sample, so there will be more variation from one sample to the next than if we take very large samples. Now look again at the formula:

$$s.e. = \frac{\sigma}{\sqrt{n}}$$

The reasons for the exact mathematics are complex, but it's clear that a larger population standard deviation increases the size of *s.e.*, and a larger sample decreases the size. Now if we calculate

$$s.e. = \frac{\sigma}{\sqrt{n}} = \frac{2}{\sqrt{25}} = \frac{2}{5} = .4$$

we find that $s.e. = .4$ ounces.

What does this tell us about the accuracy of our samples? Since we have the means and standard deviations, we can graph the population and sampling distributions this way:

Figure 4

a. Population Distribution

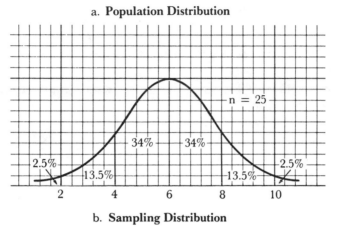

b. Sampling Distribution

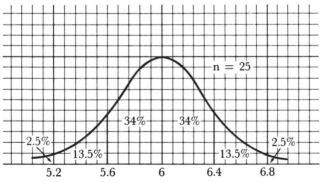

We see in Figure 4b that 68% of the samples will have means within .4 ounces of the true mean of the population. Now we can see why we use the term "standard error" and the abbreviation $s.e.$ — it gives us a sense of how big or small an error we are likely to be making when we rely on a sample to tell us about the whole population. The larger the spread of the sampling distribution, the more difference there is likely to be between the mean of any particular sample we take and the mean of the population. In the following two chapters, we will examine specific uses of this concept in statistical inference.

Exercise: Use Example 8 substituting $\mu = 5$ oz., $\sigma = 1.5$ oz. and $n = 16$ and graph the population distribution and sampling distribution. What percent of the samples will have means of 4 ounces or more? Use z-scores.

We must note one more point about the sampling distribution of the mean before moving on. For our formulas to work, the sampling distribution must be a normal one. But is it always safe to assume that it will be? A key characteristic of the sampling distribution of the mean is that it tends to be normal even if the distribution of the whole population from which the samples come is not normal. Why? The whole proof is tedious, but remember that in any non-normal distribution, there are only a very few individuals with very low or very high scores who cause it to be skewed; when we're dealing with sample means, these few extreme individuals tend to be outweighed by the mass of others in the sample. Assuming the sampling distribution of the mean to be normal is likely to get us into trouble only when the size of the sample is very small.

Binomial Sampling Distribution. We have seen how to estimate errors in the case of the mean of a sample. Now we will look at the sampling distribution for another important statistic, the binomial probability, p. We have already encountered the binomial distribution in Chapter 8, where we saw that it was useful in describing events which had two mutually exclusive possible outcomes. The binomial sampling distribution will help us to estimate the possible error when we use a sample to predict the probability of one of the outcomes of a binomial experiment occurring in a given population.

Just as in the case of the sampling distribution of the mean, or any other normal distribution, all we need to describe the binomial sampling distribution is its mean and its standard deviation.

1. The mean of the sampling distribution is p, the binomial probability for the population, just as the mean of the sampling distribution of the mean is the mean of the population.

2. The standard error, or standard deviation of the sampling distribution is given by the formula:

$$s.e. = \sqrt{\frac{p(1-p)}{n}}.$$

Notice again that the larger the sample, the smaller the standard error will be.

Example 11: Suppose that the orchard in Example 8 produces equal numbers of red and golden delicious apples. We again take samples of 25 apples each. But this time, instead of being concerned with the mean weight of the apples in the sam-

ple, we are interested in the probability that an apple selected at random from a sample will be red. We know that
an apple selected at random from the orchard has a 50%
probability of being red. The sampling distribution will
give us the probabilities for the samples.

We know from the characteristics described above that
the mean of the distribution will be $p = .5$ and the standard
error will be

$$s.e. = \sqrt{\frac{(.5)(1-.5)}{25}} = \sqrt{\frac{.25}{25}} = \frac{.5}{5} = .1$$

Figure 5

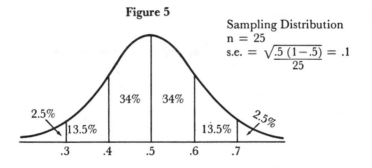

Sampling Distribution
n = 25
$$s.e. = \sqrt{\frac{.5\ (1-.5)}{25}} = .1$$

Therefore we see that in 68% of the samples, the probability of picking a red apple will be between .4 and .6. In 13½% of all possible samples it
will be between .3 and .4. In 2½% of the cases, the probability of getting a
red apple will be .7 or better, and so on.

PART B: The Perils of Probability

How often have you heard the argument: "The probability of this
happening by chance is one in a million, so it must have happened because . . ." Many different assumptions go into an argument of this sort
and until we know what they are, we can't decide on the validity of the
argument. The assumptions may involve a choice of formula, a choice of
numbers to substitute into the formulas, or the interpretation of the results
of the calculation.

Assessing Assumptions

We will begin by considering an example in which a number of statisticians engaged in a public debate about the assumptions involved in deciding the significance of a certain event. The event was the deaths of three
people after inoculation with a flu vaccine in a Pittsburgh clinic. Because
of the potential effect on the program of inoculations, the incident quickly

became the subject of much speculation and considerable controversy.

Example 12: An article by Philip M. Boffey in *Science* (November 5, 1976) begins by setting forth the problem:

> The odds against it happening by chance seemed enormous. Three elderly individuals went to the same public health clinic here to receive influenza shots on October 11. They were inoculated within an hour of each other from the same lot of vaccine. And all died from 1 to 6 hours later. (pg. 590)

The question, then, is whether the three deaths occurred by chance or not, and there was considerable disagreement about the answer:

> The disagreement stems primarily from different assumptions about the appropriate population base to be considered in estimating probabilities. . . . Does one ask how likely it is that three deaths would occur normally on a given day among the relative handful of people who were vaccinated during the critical hour at that one clinic, or among the 1200 who were vaccinated at the clinic that day, or among the 8000 or so elderly who were immunized in Allegheny County that day or among even larger groups of elderly immunized in the state or nation?
>
> Using one set of assumptions, the Center for Disease Control, which is promoting the immunization campaign has managed to calculate the odds as high as 1 in 50 that the deaths would occur normally. Using another set, the county coroner's office, acting as devil's advocate, puts the odds as low as 1 in a million. One neutral expert — Robert J. Armstrong, chief of mortality statistics at the National Center for Health Statistics — has a gut feeling that the deaths were "an extremely rare event — a tremendous long shot." But he notes that highly improbable events do in fact occur. (pg. 648)

So who is right? Were the deaths accidental or were they related to the use of the vaccine? Obviously we will never know. But there are two points to be made. The first is that any calculation of probabilities rests on a set of assumptions and that these assumptions need to be carefully stated before calculations are made or accepted. The second point deals with the problem of how we assess these assumptions. What *were* the correct numbers to use in the calculation? One approach is suggested in a letter by Kac and Rubinow in the April 29, 1977 issue of *Science*. Their calculations lead them to the conclusion that the probability of the death's occurring by coincidence is about 1 in 500,000. Then they go on to say:

> It all boils down to a definition of coincidence, and our calculation shows that the result depends very sensitively on the definition. Our calculation can be looked upon as a simple statistical test of the hypothesis that the batch of vaccine used during the crucial hour was faulty. (pg. 480)

Here then is one answer to the puzzle of which assumptions are legitimate. In some situations, it makes a great deal more sense to test a particular hypothesis than to calculate the probability that the event occurred by chance. So if we suspect that one batch of the vaccine was faulty, we base our assumed values on that hypothesis — that is, the base population we use for our probability calculations will be the population vaccinated with that batch. If we suspect that the whole vaccination program is dangerous, we must use as our base population everyone who was vaccinated. There are any number of possible hypotheses that might be tested, some much more plausible than others. The more factors in the situation a given hypothesis takes into account the better. Sometimes we may want to test several hypotheses. But the point is that we must be clear about what we are assuming and why.

Unfortunately, figuring the probabilities on the basis of a particular hypothesis or explanation does not solve all problems of faulty assumptions. Here is an example where the assumptions used to calculate the probabilities are not in question, but the interpretation of these probabilities as supporting a particular explanation are.

Example 13: The January 13, 1978 issue of *Science* contains an article about the effects of a chemical plant on the health of nearby inhabitants. Data was collected and it was found that there were 8 cases of cancer among 120 people who had lived in the area for 6 or more years. The investigator, Dr. Capurro, wrote:

> "In a community where the residents were chronically exposed to a mixture of solvent vapors in the atmosphere, the annual death rate due to malignancies appears to be about seven times the rate otherwise expected."

In this case, no one argued that the deaths were actually due to chance, or that the probabilities were inaccurate. The problem here is that only one alternative to chance was proposed — the effect on residents of the chemical factory. The case went to court and at the trial another possibility was suggested: an environmental epidemiologist, William P. Radford, argued that it takes at least 10 years for this type of cancer to develop. The chemical company had been in full operation for only 2 or 3 years when the cancers appeared. "If, in fact the lymphomas (cancers) were caused by an agent in the environment," he added, "the most likely explanation was the paper mill that had operated on the . . . site previously." The argument then continued between the two parties about how long it actually does take for this type of cancer to develop.

It should be obvious from the above two examples that using probabilities to assign causes to events is a risky business at best, and yet it is quite necessary. We will have a good deal more to say on this subject in the chapter on correlations.

Another type of assumption that we see a great deal of is one that is crucial to many animal experiments.

Example 14: In the spring of 1977 the newspapers were full of stories on the controversy over whether or not saccharin should be banned as a health hazard. Data from a Canadian study suggested it might cause cancer. The *Boston Globe* of April 10, 1977 gave both sides of the argument. One side distrusted the results because:

> The quantities of saccharin used were so great that in order to consume an equivalent amount, it is said that a person would need to drink daily, for lifetime, 800 or more 12-ounce bottles or cans of diet soda. (pg. A1)

The other side countered with the following argument:

> In the human population, very large numbers of people are exposed to low doses of chemicals, but the impact of seemingly low doses of a carcinogen may not be low at all. Exposure of 200 million Americans to doses that cause one cancer in every 10,000 people, for example, would result in 20,000 cancers — clearly a public-health disaster.

> To detect the effect of low doses of a chemical that causes one tumor in every 10,000 exposed rats would require using hundreds of thousands of rats. Such vast experiments would be unwieldly and prohibitively expensive. But a dose 5000 times higher is likely to cause cancer in about 5000 of every 10,000 rats, or 50 percent.

> Administration of high dosages permits the cancer-causing effect to be readily apparent in a practical, manageable number of animals (30 to 50). The technique is routinely used in carcinogenesis tests. (pg. A1, A2)

The assumption, then, is that if a small amount of exposure causes cancer in a small number of subjects, a large amount will cause cancer in a large number. There is some empirical justification for using this technique. For example, it is known that substances are either carcinogenic or not. As the *Globe* points out, "the argument that anything can cause cancer if given in large enough doses is false. High doses of normally safe chemicals may be toxic, but they will not cause tumors." Therefore if a substance is shown to act as a cause of cancer in any given dose, it can justifiably be classified as a potential carcinogen. The evidence tells us nothing, however, about the size of the dosage that would constitute a danger for humans, since little is known of the relative sensitivities of rats and humans. This is not a problem stemming from the nature of statistics, but one stemming from respective natures of rats and humans.

Non-Independent Events

In Part A we pointed out that probabilities must be calculated differently for independent and non-independent events. We gave one example of non-independent events — the case in which three cards are drawn from a deck without replacing the card after each draw. We saw that this changed the probability of success on each draw. This principle becomes important when we draw several samples from a fairly small population. Can you see that failure to replace after each sample is drawn will affect the randomness of the samples? In very large populations, replacement becomes much less important, and for all practical purposes unnecessary.

There are a number of other ways in which events fail to be independent that have nothing to do with replacement. Some events are non-independent for theoretical and others for empirical reasons. An example of empirical non-independence is the case where a student takes the same test twice. The events cannot be independent because experience tells us that the student is likely to learn something the first time and so get a better score the second. An example of a case in which statistical theory tells us that the two events are not independent is shown below.

Example 15: Suppose we are selecting a person at random from the employees of a particular company. We know that 50% of the people working for this company are women, and 50% are secretaries. What is the probability of picking a person who is both a woman and a secretary? If sex and occupation were independent events (like, say, lefthandedness and occupation), we could multiply the probabilities. We would get a result of .25. But in fact women are much more likely than men to be secretaries, and secretaries are much more likely to be women than to be men. So the events of picking a woman and picking a secretary are not independent, and in fact the probability of picking a female secretary is much greater than .25.

One final example will show the importance of the concept of independence.

Example 16: The question of whether certain particular human characteristics are inherited or acquired from the environment, and to what extent, has been a favorite one for psychologists and others for a long time. It is a difficult subject to study since people who share the same heredity (brothers and sisters — particularly identical twins) also tend to share the same environment, the family home. Much of the research in this area has therefore concentrated on identical twins (since they share the exact same heredity) who have been separated and raised in different environments. The assumption has been that heredity and environment in these cases become independent factors. Recent work has shown this assumption to be incorrect more often than not. The following is quoted from *Shaping the American Educational State* by Clarence J. Karier. It refers to the Shields study of 40 pairs of identical twins.

". . . in 27 cases, the two separated twins were reared in related branches of parents' families; only in 13 cases were the twins reared in unrelated families. . . . The typical case of 'unrelated families' was one in which the mother kept one twin, and gave the other to friends of the family." (pg. 384)

We must be very careful, whenever we use probability estimates for complex events, that we know whether the events are independent or not. It is only when we are sure of their independence that we can mechanically apply the simple formula for independent events.

The Individual and the Distribution

One final kind of reasoning error with probability involves the notion of independence in a very different sense from that used above. In this case what happens is that we look at one individual instance independently of the distribution from which it is drawn. The idea is best explained by the following example.

Example 17: On February 7, 1978, the *National Enquirer*'s headline read: "AMAZING WINNER OF ENQUIRER'S ESP TEST: 5 OUT OF 5 PSYCHIC PREDICTIONS CAME TRUE." The story on pg. 37 begins: "In a stunning display of psychic ability, Mrs. Florentine Von Rad-Keyye Kaiser scored hits in five predictions of 1977 events." The implication is that scoring 5 out of 5 predictions correct is so improbable that it must mean real psychic powers.

Let's examine the situation. Just how unlikely it is depends of course on the probability of each individual prediction. Predications that "the arms race with Russia will continue" and that "There will be lots of action on drugs . . . on the Mexican border" seem fairly sure bets. Predicting the deaths of Charlie Chaplin and Groucho Marx was taking somewhat more of a chance, but none of the predictions were highly unlikely. If we assigned an overall probability of .25 for each prediction, it would definitely be giving the woman the benefit of the doubt. The overall probability of her scoring 5 correct predictions is then .25 multiplied by itself 5 times or approximately .001. That means the probability of doing this well by chance is about one out of a thousand. But wait a minute. This was part of the contest. There were in fact 1500 entries. Since the probability of getting five correct answers is .001 (one out of a thousand), 1500 entries can easily yield one "amazing winner."

Our knowledge of how to calculate independent compound probabilities got us that far. But suppose we want to know exactly how good the chances of those 1500 entries producing a winner were. We can answer this question through the use of sampling distributions. What we have is a binomial distribution in which the two either-or categories are having five correct answers and not having five correct answers. Our desired event is *having* the answers, and we know that for this event $p = .001$. The sample is the 1500 people who responded to the "ESP test," so $n = 1500$. We calculate the mean and the standard deviation, and the resulting graph looks like this:

Figure 6

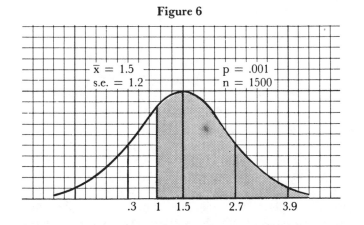

The shaded area in the graph represents the probability that one or more of the 1500 contestants will get 5 out of 5 predictions right. Since 1 is .42 standard deviations to the left of the mean, we look up .42 on the table of z-scores and find that the probability of one of the contestant's making a perfect score is 66%.

SUMMARY

PART A

Probability is the branch of mathematics devoted to the study of chance. An **event** is the outcome of an experiment or an observation which can have two or more possible outcomes. We use the **relative frequency concept** of probability to calculate the probability of any particular event occurring. In the case of **theoretical probabilities,** those in which we know in advance all possible outcomes and their frequencies, we find the probability of an event by dividing the number of possible ways in which the event can occur by the total number of ways in which all possible outcomes of the experiment can occur. In the case of **empirical probabilities,** we perform the experiment a great many times and divide the frequency of the event by the number of times the experiment was performed. The formula for finding probabilities of simple events can then be written:

$$p(e) = \frac{n(e)}{n(t)}$$

where $p(e)$ means the probability of the event, $n(e)$ means the number of occurrences of the event, either theoretically or empirically determined, and $n(t)$ means the total number of ways all possible outcomes can occur (theoretically) or the total number of experiments (empirically).

A **simple event** is one that meets only one condition, while a **compound event** must meet two or more conditions. The simple events making up a compound event may be either **independent** or **non-independent.** Two events are independent when the occurrence of one has no effect on the probability of the occurrence of the other. If the events making up a compound event are independent, the probability of the compound event can be calculated by simply multiplying together the probabilities of the simple events that must occur for the compound event to happen. If the simple events are not independent, this formula will not work.

A **probability distribution** is a distribution showing the probabilities associated with each of the possible outcomes of an experiment or observation. Usually probability distributions are based on theoretical probabilities.

A **sampling distribution** is a probability distribution which shows the probability of getting particular samples from a given population. A **sampling distribution of the mean** is the distribution of the means we get when taking many samples of the same size from a given population. The sampling distribution is a normal distribution even when the population distribution is not. The mean of the sampling distribution is always the same as the mean of the population. The standard deviation of the sampling distribution of the mean is also called the **standard error.** It is found by dividing the square root of the size of the sample into the standard deviation of the population:

$$s.e. = \frac{\sigma}{\sqrt{n}}$$

A **binomial** sampling distribution is used to help us estimate the possible error when we use a sample to predict the probability of one of the outcomes of a binomial experiment. For a binomial distribution we use the following formula:

$$s.e. = \sqrt{\frac{p(1 - p)}{n}}.$$

p = probability of one of the two possible outcomes; the mean of the sampling distribution

n = size of sample

$s.e.$ = standard error or standard deviation of the sampling distribution

PART B

In order to apply the formulas of probability theory to any particular situation, it is necessary to make a number of assumptions. These assumptions must be explicitly stated and carefully checked before the calculated probabilities can be accepted. Sometimes it makes more sense to test a particular hypothesis than to calculate the probability that something occurred by chance. When using this method, however, it is necessary to make sure that all possible or reasonable hypotheses are tested. It is also important that the results of a probability calculation not be generalized to a situation different from the one the assumptions were made about.

When calculating probabilities of compound events it is important to consider whether the single events making up the compound event are independent or not. Some reasons why events may not be independent include:

1. Failure to replace when drawing repeated samples from the same population.
2. Making repeated experiments with the same subjects.
3. The single events constituting overlapping categories.
4. Assuming empirical independence when such is not the case.

A final error sometimes made in working with probabilities is to calculate a single probability rather than a whole probability distribution. A single event may seem very unlikely by itself, yet its occurrence may be quite probable when it is considered as one of a large number of experiments or observations.

EXERCISES

PART A

1. Assume a jar contains 25 white balls, 75 green, 50 red and 100 black balls.
   a. *If we draw one ball, what is the probability of its being red?
   b. If we draw one ball, what is the probability of its being either red or black?
   c. *If we draw two balls, what is the probability of one being red and the other white?
   d. If we draw one ball on three separate occasions, what is the probability of getting a red ball each time if we replace the ball after each draw? Will the probability be the same if we don't replace the ball?

2. Suppose we have a population of normally distributed test scores with a mean of 60 and a standard deviation of 14.
   a. Draw a graph of the population distribution.
   b. Draw a graph of the sampling distribution for the above population if the size of the samples is $n = 49$. What is the mean of the sampling distribution? What is its standard deviation?
   c. What is the probability of drawing a sample ($n = 49$) from this population with a mean of 66 or more?

3. *We are interested in an election with two candidates — Smith and Jones. If we assume that the chances of either winning are about even, what is the probability that out of a sample of 25 voters, at least 15 will vote for Smith?

4. Suppose we are investigating 10,000 students who form the entire population of a particular university. We find that their mean grade point average is 2.1 with a standard deviation of .3.
   a. Assuming a normal distribution, draw a graph for the population.
   b. Draw a graph of the sampling distribution for $n = 100$.
   c. Draw a graph of the sampling distribution for $n = 10,000$.
   d. Draw a graph of the sampling distribution for $n = 1$.

5. *A company has an equal number of male and female employees. If men and women have equal chances to be promoted, what is the probability that selecting more than 65 of the 100 executives will be male by chance?

PART B

1. Suppose the population base for the following questions is the student body of a large university. Which of the following compound events are made up of independent simple events?
   a. *Selecting a person at random from the population who is both blue-eyed and blond.

    b. Selecting a woman with blue eyes from the population at random.

    c. *Selecting a student making an A in 4 out of 4 classes during one semester.

    d. Selecting at random a student who owns a car and lives more than 5 miles from campus.

2. What is wrong with the following arguments?

    a. I've flipped this coin three times and it has come up heads each time. Since it should come up heads only half the time, I'll bet it will come up tails next time.

    b. If the chances of getting a winning lottery ticket are 1 in 1,000, I'll buy 1,000 tickets to make sure I win.

    c. An article entitled "Are Career Interests Inheritable" appeared in the March 1978 issue of *Psychology Today*. The authors say, "We found a consistently moderate but statistically significant degree of similarity between biological parents and their children. Since resemblances between adoptive parents and their children were essentially nonexistent, we concluded that the convergence of interests within the biological families could only be accounted for by some sort of biological component." (pg. 90)

    d. The probability of a new brand of pain reliever causing serious side effects in adults is 1 in 100,000. We can expect the probability to be the same for children.

3. *Review Example 12 in Part B. A letter in the March 11, 1977 issue of *Science* makes the following comment: "Probabilistic arguments show that, although the chances of three or more deaths in any one clinic in a single day are minute, the chance that some clinic would experience three or more deaths on some day during the first week of the inoculation program is appreciable and could easily be as high as 10 percent, even if the vaccine is perfectly safe." (pg. 934) Explain what the authors mean in your own words and either attack or defend their approach to the problem.

Chapter 11

HYPOTHESIS TESTING

PART A

In the last chapter, we studied the effect of chance on the data that are collected from a sample of the population. Now we are going to see what conclusions can be drawn from such data, taking into account all that we know about selecting proper samples and about how closely the sample can be expected to resemble the population from which it was drawn. In other words, we are going to look at how statistical inference is used in research. The problems involved in statistical analysis of experimental results fill volumes, if not libraries. All we will do here is give a few examples of how inferential statistics can be used.

In Chapter 12, we will discuss in detail how to measure the reliability of estimates about a whole population made on the basis of what we hope are typical samples. In this chapter our problem is somewhat different, although it requires the same basic tools. We will be dealing with samples that appear to be *atypical*. We will want to know whether they show a really significant deviation from the norm, or whether they are in fact no more unusual than many other randomly selected samples might be.

For instance, if we discover that, in a sample drawn from our city, 8% of the labor force is unemployed as compared to a national average of 7%, does this mean that our city's unemployment problem is very unusual? Or are there quite probably many equally large unemployment rates in other randomly selected cities? Making such distinctions may appear to be splitting hairs, but they turn out to be crucial in determining whether a given statistic does or doesn't offer strong support to an argument. Before we can appreciate the importance of this distinction, we must briefly look at the way formal arguments are presented in the natural and social sciences.

Stating the Hypothesis. Before we can perform an experiment or analyze it statistically, we must carefully state the question we are trying to answer. Let's look at an example.

Example 1: A doctor working in an asbestos factory suspects that the working conditions in this factory are causing respiratory illnesses among the workers. She studies a sample of 100 workers over a period of time to see if she can find evidence to support her hunch.

The question the doctor is trying to answer is whether or not working in the factory is dangerous to people's health, but in order to test this question statistically, she will have to put the question in the form of a statement that can be tested to see if it is true or not. Such a statement is called a **hypothesis.** The statistical analysis will tell us whether the hypothesis should be accepted (considered true) or rejected (considered not true). (Notice that a rejected hypothesis has not been proven false; it simply hasn't been backed up by good enough evidence for us to consider it true.)

In order to do a statistical analysis of a hypothesis, we must have some numbers to work with. In the case of the workers in the asbestos factory, we might count the number of cases of lung cancer among workers, or the number of days lost from work due to respiratory disease, or the number of cases hospitalized for such disease, or any one of a number of different measures. Suppose we decide to count the number of days lost from work among the workers in the sample. We now have one numerical measure, but that is not enough. It doesn't help to know that the average worker from this factory missed three, or five, or twenty days from work because of respiratory disease. Members of every population suffer from colds, bronchitis, even lung cancer. We need to know whether these workers suffer from such illnesses more than other populations do. So we must have something to compare them to. In this case, we would want to know how many days are lost from work because of respiratory disease among all other factory workers.

So far we have said that we must phrase our question in the form of a statement or hypothesis which can be accepted or rejected, that it must be based on numerical data and that we must be able to make a comparison. We are now ready to state our hypothesis. It will be a statement comparing the number of days of work missed by workers at the asbestos factory because of respiratory illness and the number of days missed by workers in general for such illness. There are many ways in which we could state that comparison, but basically we are interested only in whether or not there is a difference between the sample and the population at large. We can state our hypothesis either positively or negatively:

1. There is a difference between the number of days missed by the

   asbestos factory workers and all other factory workers.
2. There is no difference between the number of days missed by the
   asbestos factory workers and all other factory workers.

The first statement is the one we want to- prove — it is the one we think is correct. We can call this statement the **research hypothesis.** However, it is not the statement we will use in our statistical analysis. Instead, we will use the second statement, called the null hypothesis.

Notice the relationship between the two statements. If one is true, the other must be false, and vice versa. Between them they cover all possibilities, and they are mutually exclusive. Whenever we use statistics to test hypotheses, we always begin by writing out two such statements — one positive and one negative — that cover every possible outcome of the experiment. The **null hypothesis** is always a negative statement. It takes the form "There is *no* difference . . ." The positive statement, "There is a difference . . ." is called the **alternative hypothesis.** Usually the alternative hypothesis is the same as the research hypothesis. Occasionally, however, we want to show that there is indeed no difference between our sample and the general population. For example, a drug company may conduct research to show that people using a particular medication do not have higher blood pressure than the general population. In that case the research hypothesis would be the same as the null hypothesis. However, no matter which of the two is the research hypothesis it is always the null hypothesis that we test statistically.

If our analysis leads us to accept the null hypothesis, we must reject the alternative hypothesis, since they are always mutually exclusive. However, if we reject the null hypothesis, then we must accept the alternative hypothesis. Notice that we talk about accepting or rejecting the hypothesis rather than proving or disproving it. That is because our results can never have the logical certainty of deductive logic. Even when we are rejecting rather than accepting, we still will not get certain answers. That is because in the real world we are not dealing with absolutes. It is possible inductively to disprove statements such as "All crows are black" by coming up with a white crow, or "No elephants have wings" by finding one that does. In real life, however, we are much less concerned with the occasional aberration and much more interested in what is generally true. When we test to see if there is a difference between two sample or population measures, we will not expect to find absolute proof that there is or isn't, but we will be able to say something about how likely we are to be right or wrong when we decide to accept or reject the hypothesis.

Testing the Hypothesis. Continuing to use the situation in Example 1, we can now state the null hypothesis (H_o) and the alternative hypothesis (H_a):

H_o: There is no difference between the mean number of days missed by workers at the asbestos factory because of respiratory disease and that of all other factory workers.

H_a: There is a difference between the mean number of days missed by workers at the asbestos factory because of respiratory disease and that of all other factory workers.

Exercise: Assume a tire manufacturer states that the mean life of its tires is 65,000 miles. The population mean is known to be 64,000 miles. We want to know if the tires produced by this manufacturer are actually better. State the null and alternate hypothesis.

The numerical value (in this case a mean) that we calculate from our sample, is called the **test statistic;** the comparable measure that we know to hold for the general population is called the **population parameter.** We will need values for these before we can test the hypothesis.

Example 1: (cont.) Let's suppose that the doctor at the asbestos factory finds that the mean number of days lost from work because of respiratory illness among her sample of 100 workers was two days during the last year. So the test statistic is $\bar{x} = 2$. Let us further suppose that the comparable figure for all other factory workers has a mean of 1.5 with a standard deviation of .8. Then the population parameter will be $\mu = 1.5$. We can now state our null hypothesis:

H_o: There is no difference between $\bar{x}$ and μ.

Stated in this way, the null hypothesis seems obviously false. Since $\bar{x}$ = 2 and μ = 1.5, there seems to be a clear difference between them. But how reliable was the doctor's sample? We know from the last chapter that a sample quite often does not have the same mean as the population it comes from, even if the sample is selected randomly. Perhaps the higher mean absence rate of the asbestos workers is just a matter of chance, the sort of difference we might easily run into if we compared any sample of only 100 workers to the mean for all other workers. If this were true, we would have to say that we had no "statistically significant" evidence that the asbestos workers missed any more days, on the average, than other workers.

What we need to know, then, is whether the difference between the test statistic and the population parameter is large enough for us to suspect that the sample is not one we would be likely to get through random sampling of the general population. In order to find out, we will use the techniques developed in the last chapter. Remember that there we used the notion of a sampling distribution to help us predict the likelihood of getting a certain mean for a sample of a given size if we knew the mean and standard deviation of the population from which the sample was taken. We now have a sample in which $n = 100$, drawn from a population with a mean of 1.5 lost days and a standard deviation of .8. What is the likelihood of drawing a random sample with a mean of 2.0 days lost?

The sampling distribution is given in Figure 1, with the arrow showing the location of the test statistic.

Figure 1

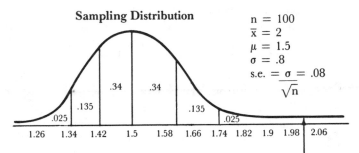

Recall that the probability of a sample mean falling within a given class corresponds to the area under the curve of that class. So 68% of all samples will be in the two classes that are within one standard deviation of the mean, 95% within two standard deviations, and so on. We see then that the probability of getting a sample of 100 from the general population with a mean of 2 or more is close to 0. We therefore reject the null hypothesis that there is no difference between the asbestos factory workers and all other factory workers and accept the alternative (and research) hypothesis that there is a difference.

This case was very clearcut and we had no difficulty in seeing that we had to reject the null hypothesis and therefore accept the alternative or research hypothesis. But suppose the mean of our sample had been closer to the mean of the general population. If the test statistic had been 1.66 instead of 2, the chance that the sample could have been drawn at random from the general population — that is, that there was no significant difference between the two means — would have been .025; if the test statistic had been 1.58, the probability of there being no difference would have been .16, and so on. Where is the cutoff point at which we accept or reject the hypothesis?

Significance Level. One way of looking at the problem of finding a cutoff point for accepting or rejecting a null hypothesis is to consider how large the difference is between the population mean (or whatever other measure of the population we are using) and that of the sample. In Example 1, that difference was .5 lost work days. But how large a difference is a significant one? And how can we make up a rule that can be used with a variety of different measures?

The solution is to consider significance in terms of probabilities. Look at Figure 1. We have a 2.5% probability of getting a sample with a mean of 1.66 or more lost workdays by chance. We have a 0.15% probability of getting a sample with a mean of 1.74 or higher, and so on. We will decide what constitutes a significant difference, then, not on the basis of the actual numbers we get, but on the basis of the probability, calculated from the sampling distribution, that we could get the sample mean by chance from a population with the given mean. This probability is called the **significance level.** It measures the probability that the null hypothesis is actually true. If we reject the null hypothesis, then the significance level is the probability that we are in error.

The significance level can be used in two ways: to set a cutoff point before we carry out our experiment or observation, and to measure the probability of error after we have reached a conclusion.

1. Cutoff point: we always set a necessary significance level beforehand, so that it will be clear what values of the test statistic will lead to acceptance or rejection of the null hypothesis. Since the significance level measures a probability, we will use the letter P to designate it, using the capital to differentiate it from the small p we have used for other types of probabilities. Suppose that in Example 1 we had chosen a significance level of $P = .025$. We want a probability of no more than 2.5% that we are mistakenly rejecting the null hypothesis. What is our cutoff point? What mean number of lost workdays in our sample has at most a probability of .025 of occurring by chance? From our sampling distribution we know that this is the sample mean two standard deviations distant from the population mean. Since we are concerned in this experiment with a sample mean *greater* than the population mean, we move to the *right*. The value of the cutoff point (which in this case is 1.66 lost workdays) is known as the **critical value.** If the test statistic falls between the critical value and the mean, the null hypothesis is accepted; if it falls beyond the critical value, then the null hypothesis is rejected. In Figure 2 below the shaded portion of the graph shows the rejection region for a significance level of $P = .025$. The **rejection region** is the part of the graph which tells us to reject the null hypothesis if the test statistic falls within it, because the probability of getting this value by chance rather than because there is a real difference is so slight.

We chose the significance level of .025 because it was an easy one to demonstrate from the graph without resorting to the use of z-scores. In the social sciences it is customary in most cases to set the significance level at .05. Of course, the smaller the value of the significance level, the smaller the chance that one is making an error in rejecting the null hypothesis.

Figure 2

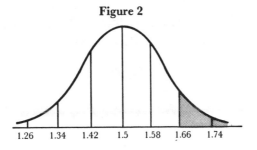

1.26   1.34   1.42   1.5   1.58   1.66   1.74

Stating that same point without all the double negatives, the smaller the number we choose as P, the greater the chance that, if we determine the null hypothesis false, our conclusion will reflect reality rather than a statistical fluke.

2. After the fact: even though the significance level is set in advance to decide when to accept or reject the hypothesis, it can also be calculated after the experiment has been conducted to show just how significant the results are. Suppose you decide that in a certain case you will reject the null hypothesis at a significance level of .05, but after you perform your experiment you find that you actually have a significance level of .001. Since the significance level measures the probability of error, the smaller the significance level, the more probable it is that you are right. It is therefore a good idea to report the actual probability if it is smaller than the previously set significance level, and many investigators now do that.

We have then, three interrelated concepts connected with hypothesis testing when a comparison is being made between means: the **significance level,** the **critical value,** and the **rejection region,** and we need to be clear about how these are obtained.

1. The significance level is set before any experiments or observations are performed, by the experimenter on the basis of past experience, convention, etc.

2. The critical value is calculated from the sampling distribution and the significance level. In the above example we used a significance level of .025 because that required little calculation. We could see by inspection that if we wanted to have a probability of .025, we needed to move two standard deviations from the mean, so the critical value was $1.5 + 2(.08)$ = 1.66. If we had a more usual significance value, such as $P = .05$, we would have to use the z-score table to find the critical value. In Chapter 8 we calculated the z-score and then looked on the table to find the area that corresponded to it. Now we will use the table in the opposite direction. We need to know the z-score corresponding to a certain probability — that is, to a certain portion of the area under the curve. Remember that the area

on the table is the area between the mean and the z-score. So the area we look up on the table when we want a significance level of P = .05 is .45, not .05. Figure 3 below should make clear why we must do this.

Figure 3

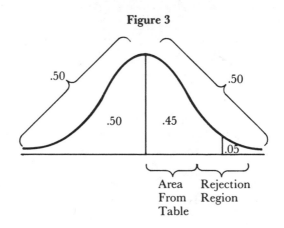

When we look up .45 in the area column we find that it is approximately equal to a z-score of 1.65. That means that the z-score of the critical value we are looking for is 1.65, or that the critical value is 1.65 standard deviations to the right of the mean. We must therefore add 1.65 standard deviations to the mean: $1.5 + 1.65(.08) = 1.632$.

Notice that the significance level of any particular test statistic can be calculated by performing the same operations in reverse. Suppose we have a test statistic of 1.6 and we want to know what P is. We convert 1.6 to a z score:

$$z = \frac{1.6 - 1.5}{.08} = 1.25$$

We then look up 1.25 on the area table and find that the corresponding area is .3944. Since that is the area between the mean and the critical value, and we want the area on the other side of the critical value, we subtract from .5000 and get .1056. The significance level associated with 1.6 lost workdays is therefore slightly more than .1.

3. The rejection region is simply the area of the graph on the other side of the critical value from the mean.

One- and Two-Tailed Tests. When we talk about a cutoff point or critical value for accepting or rejecting the null hypothesis in the above example, we are talking about only one point. If we are looking at the health of workers in an asbestos factory, we are interested in whether or not their health is worse than the average. No one has suggested that working in such a factory can actually be good for people, so we need not consider the possibility that these workers lose fewer days from work than all other factory workers. The cutoff point will be to the right of the mean and we must decide how far to the right. If we are looking for only one cutoff point, we are using what is called a **one-tailed test.** Often, however, we have no way of knowing in advance whether our sample will differ from the population by having a higher or a lower mean. For example, we may be testing a new method for teaching statistics. We cannot know if it will be more effective — it may actually be less effective than the old method. Therefore if we test the progress of students using the new method against that of students using conventional methods, we must establish two cutoff points — one to indicate that the method is different by being better and one that it is different by being worse. When we have two cutoff points, one on either side of the mean, we say we are using a **two-tailed test.**

Since we want to set the significance level independently of whether we are conducting a one- or a two-tailed test, we set the level first, and then determine our rejection regions on the basis of whether we want one or two. The graphs in Figure 4 below show examples of all possible rejection regions when the significance level is .05, where 4a is a two-tailed test and 4b and 4c are one-tailed tests. Note that in all three cases the total area of the rejection region is the same; thus in the two-tailed test we use a different critical value, so as to produce two rejection regions, each with an area (probability) of .025.

Figure 4

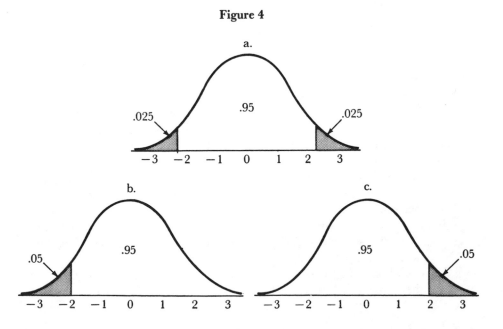

Type I and Type II Errors. Since the probability we use as the significance level tells us what the chances are of being in error when we reject the null hypothesis, we might think that it is desirable to set the significance level to the smallest possible number. Why use a significance level of .05 and risk a 5% chance of being wrong if we are testing the effectiveness of a new drug, for instance? Why not set it at .01 and have only a 1% chance of being wrong? The trouble with such reasoning is that there are actually two types of errors we can make. A **Type I error** is the error of rejecting a true null hypothesis. That is, if our null hypothesis is that the drug makes no difference in the treatment of the disease, and we reject this hypothesis at a .05 level, there is a 5% chance that we are wrong, and the drug is not effective. But if we set the significance level at .01, we increase the danger of making a Type II error. A **Type II error** is the error of accepting a false null hypothesis. With a significance level of .01 we are more likely to overlook an effective drug by accepting the hypothesis that it makes no difference in treating the disease. So we see that the smaller our chance of making one of these errors, the larger our chance of making the other.

Let's look again at the case of the factory workers to see how these two types of errors are related. It is clear from the graphs below (Figure 5) that the farther out we move the rejection region, the less likely we are to make the error of rejecting the null hypothesis when it is actually true. Suppose the null hypothesis is true. If we use a critical value of 1.58 as the cutoff point, we reject the hypothesis if the test statistic is more than one standard deviation above the mean and there is a .16 chance of making an error. If we set the critical value at 1.66, the probability of error is .025, and at 1.74 it is only .0015. But let's suppose that the null hypothesis is, in fact, false. Notice that as we move the rejection region farther out, we are increasing the chance of making the error of accepting the false null hypothesis.

On the one hand, we must guard against being too easily satisfied and accepting the null hypothesis on too little evidence. On the other, we must be sure we are not asking for too much, and so running the risk of rejecting a null hypothesis that is true. As we have said before, there is no hard and fast rule about where we set the significance level, though .05 is customary in the social sciences. We can now see, however, why there might be occasions where we would use other values. For example, we might be testing a cold-relieving drug with potentially dangerous side effects. We examine these side effects with a null hypothesis stating that the number of deaths among those using the drug is no different than the number of deaths in the general population. We want to be sure that we were not accepting the hypothesis when in fact it was false. That is, we would want to avoid, as

Figure 5

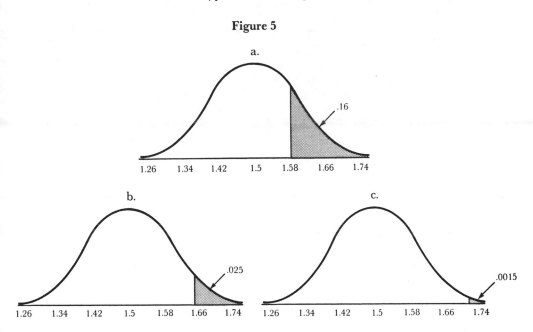

much as possible, making a Type II error, and so would use a relatively large number for our significance level. If, alternatively, we were looking at a drug that might kill cancer cells but was also dangerous to use, we would want to be very certain that it was effective before we exposed people to the dangers. Therefore we use a null hypothesis about the effectiveness of the drug. We want to see a substantial difference in the number of cancer cells destroyed by the drug and the number destroyed without it. We would want to avoid making a Type I error, and so would set our significance level to a very small value. Of course we never know whether we are making an error or not, but we can decide which of the two kinds of error we are less willing to make and then calculate the relevant probability. It should be pointed out that calculating the probability of making a Type II error is beyond the scope of this book. However, the probability of making a Type I error is just the significance level, and we know that the smaller the chance of making that error, the larger the risk of making the other.

The One-Sample Means Test

We are now ready to put all these factors together and look at an example of the kind of test we have been talking about — the one-sample means test where we test the mean of a sample against that of a population. (A two-sample means test is one where we compare the means of two different samples.) First we will list all the necessary steps and then we will apply them to a specific example.

Steps for One-Sample Means Test.

1. State the null and alternative hypotheses.
2. Decide on a significance level.
3. Decide whether to use a one- or two-tailed test.
4. On the basis of #2 and #3 above, calculate the z-score of the cut-off point. (A graph of the sampling distribution is useful here.)
5. Calculate the critical value from the z-score and the standard error. (If using a graph, indicate the rejection region on the graph.)
6. Base your conclusion on a comparison of the test statistic and the critical value. If the test statistic falls between the mean and the critical value, accept the null hypothesis. Otherwise reject it. (If using a graph, see if the test statistic falls in the rejection region or not.)

Example 2: A sample of 49 supermarkets in Boston shows that the mean price for a standard "market basket" of groceries is $37.00. If the national average for the same "market basket" is $35.50, with a standard deviation of $2, is it really more expensive to eat in Boston?

1. **Hypotheses.** H_o: There is no difference between the mean cost of a standard "market basket" of groceries in Boston and in the nation.

           H_a: There is a difference between the mean cost of a standard "market basket" of groceries in Boston and in the nation.

2. **Significance Level.** Set $P = .05$

3. **Two-Tailed Test.** We use a two-tailed test because there is no way of knowing in principle whether Boston is more expensive or not.

4. **Z-Score.** Since we are using a two-tailed test, we must divide the .05 significance level between the two tails. The area we want at each end is therefore .025. We could look up .475 (that is, $.500 - .025$) on the area part of the Z table, but probably we don't need to since we remember that that area corresponds to 2 standard deviations from the mean. So $z = 2$.

5. **Critical Value.** We can just look at the graph below to see that the relevant critical value is $36.08. Or we can calculate both critical values from the mean and the standard error:

$35.50 \pm 2(\$.29) = \$35.50 + \$.58$ and
$35.50 - \$.58$.

So the two critical values are $36.08 and $34.92.

6. Conclusion.   Since the test statistic is $37, and so is larger than the critical value, it lies in the rejection region and we reject the null hypothesis. That is, there is a difference between the mean cost of a standard "market basket" of groceries in Boston and in the nation.

Figure 6

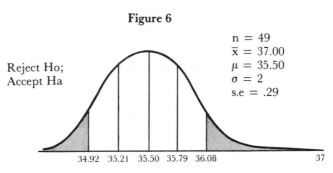

Reject Ho;
Accept Ha

n = 49
x̄ = 37.00
μ = 35.50
σ = 2
s.e = .29

34.92   35.21   35.50   35.79   36.08   37

Exercise: Based on the data in the previous exercise page 216, use the one-sample means test to test the null hypothesis.

T-Tests

Now that we understand the basic ideas behind hypothesis testing, we can look at one or two variations on that theme. First we will consider t-tests, because they are so widely used. T-tests are used with data that comes from small samples — usually samples where *n* is less than 30. T-tests are also used when we do not know the standard deviation of the population from which the sample is drawn. Remember that in the one-sample means test for large samples, we calculated the standard error in order to find the critical value. The formula we have used so far to calculate the standard error has been:

$$s.e. = \frac{\sigma}{\sqrt{n}}$$

If we do not know the value of σ, the population's standard deviation, we cannot use this method to calculate the standard error. What we can do, however, is to approximate the population's standard deviation by using the standard deviation of the sample. We then have:

$$s.e. = \frac{s}{\sqrt{n}}$$

This approximation of the population standard deviation, together with a small sample size, introduces a greater probability of error so that the normal distribution no longer works as a model for our hypothesis test. We therefore use what is known as a t-distribution. (In a case where the sample is small but the population standard deviation *is* known, we can use our old formula for *s.e.,* but we must still follow t-distribution procedures for evaluating it.)

There are actually many t-distributions, one for each sample size. Each is bell shaped and symmetric, like a normal distribution, but the relationship between the number of standard deviations from the mean and the areas under the curve are different. As the sample size gets bigger, the t-distribution becomes more like the normal distribution, so that by the time $n = 30$, the difference between the t-distribution and the normal distribution becomes insignificant and we again use the normal distribution. Below is a comparison of the normal distribution with various t-distributions.

Figure 7

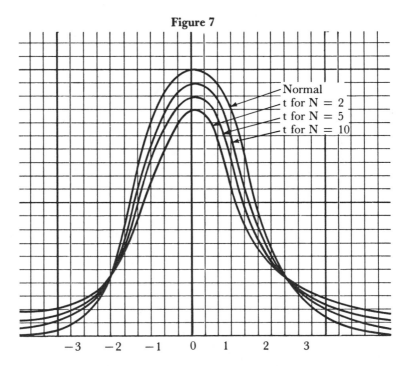

We can use the t-distribution to compare a sample mean to a population mean, just as we used the normal distribution with a larger sample. Just as previously we used a z-score to find the critical value for the appropriate significance level, now we will use a t-score. But because there are many different t-distributions, one for each sample size, the table of t-scores is arranged somewhat differently. When we used the table of z-score values, we had one piece of information (either the z-score or the area) and looked up the other. In the case of the table of t-scores we will need two pieces of information (either the sample size and the t-score or the sample size and the area) in order to find the third. Except for these minor differences, the procedure for doing a t-test is exactly like that for a large sample test of the mean.

Steps for One-Sample Means Test Using T-Distribution.

1. State the null and alternative hypotheses.
2. Decide on a significance level.
3. Decide whether to use a one- or two-tailed test.
4. Look up the t-score on the table of the t-distribution, using appropriate a and n.
5. Calculate the critical value from the t-score and the standard error.
6. Base your conclusion on a comparison of the test statistic and the critical value. If the test statistic falls between the mean and the critical value, accept the null hypothesis. Otherwise reject it.

Example 3: Suppose a certain bank is suspected of discriminating against women by failing to promote them beyond the level of tellers. The bank claims to have a successful affirmative action program. A sample of 25 of the bank's branch offices shows that there are a mean of 2.7 women in management positions per branch, with a standard deviation of 1. The mean number of women in management positions in all banks is 3 per branch. Is the bank discriminating against women compared to other banks?

1. **Hypotheses.** H_o: There is no difference between the mean number of women in management positions of this bank and of all other banks.

   H_a: There is a difference between the mean number of women in management positions of this bank and of all other banks.

2. **Significance Level.** Set P = .05.

3. **Two-Tailed Test.** We use a two-tailed test because there is no way of knowing in principle whether the bank promotes more or fewer women than other banks.

4. **T-Score.** Since we are using a two-tailed test we must divide the .05 significance level between the two tails. The area we want at each tail then is .025. So a = .025 and n = 25. Looking at the table, (in text, p. 306) we find t = 2.064.

5. **Critical Value.**   The table tells us that the t-score of the critical value is 2.064. The standard error is now calculated by the formula $s.e. = \frac{s}{\sqrt{n}}$, so $s.e. = \frac{1}{\sqrt{25}} = \frac{1}{5} = .2$.

The critical value, then, is the mean, plus the distance from the mean, or $3 \pm .2(2.064) = 3 + .4128$ and $3 - .4128$. The two critical values, then, are 3.4128 and 2.5872.

6. **Conclusion.**   Since the test statistic is 2.7, and since that number lies between the critical value of 2.5872 and the mean of 3, we accept the null hypothesis, that is, there is no difference between the mean number of women in management positions of this bank and of all other banks.

Chi-Square

The other type of hypothesis test we will look at is the chi-square. This gets its name from the Greek letter chi which is pronounced "ki" and written like this: χ. The test allows us to compare whole frequency distributions rather than comparing one single measure like the mean. It has many different uses, in many kinds of circumstances. We will only look at one application here, testing a sample to see if its distribution matches that of the population it comes from.

Example 4:   Suppose we draw a sample from the 1,000 students at a small college. We know that the age distribution for the whole college looks like this:

Age:	under 18	18–22	over 22
Frequency:	50	750	200

We take a sample of 100 students and get a distribution like this:

Age:	under 18	18–22	over 22
Frequency:	7	70	23

We know that to perfectly match the population we should have gotten 5 under-18-year-olds, 75 in the 18–22 group and 20 over-22-year-olds. But we wonder whether the difference is significant. We can use a chi-square test to find out.

As in other hypothesis tests, we need to state the hypothesis, decide on the significance level, find a critical value, and then compare that value to the test statistic. So the basic plan is the same, but parts of the procedure are quite different. In the present example, we begin by stating the null hypothesis and then the alternative hypothesis:

H_o: There is no difference between the age distribution in the population and in the sample.

H_a: There is a difference between the age distribution in the population and in the sample.

We can again use a significance level of .05. Notice that we do not have to decide on a one- or two-tailed test since chi-squares only test the amount of difference, not the direction. Our next step, then, is to calculate the test statistic, χ^2. Since the purpose of this test is to compare two distributions, the test statistic, **chi-square,** is a measure of the difference between the two distributions. The formula looks like this:

$$\chi^2 = \Sigma \frac{(O - E)^2}{E}$$

When we compare two frequency distributions, we think of one as the observed distribution (O). In our example it would be the sample distribution. The other is the expected distribution (E), which in this case is the distribution for the college population. However, since there are 1000 people in the college and only 100 in the sample, we must divide the college distribution by 10 to get the right size to compare with our sample. We will therefore use the two distributions:

	under 18	18–22	over 22
O:	7	70	23
E:	5	75	20

We look at each class of the distribution and begin by taking the difference between the observed and expected values [$(O - E)$], then square each difference [$(O - E)^2$], divide each squared difference by the expected value $\frac{[(O - E)^2]}{E}$ and finally add up the results for each class of the distribution $\left[\Sigma \frac{(O - E)^2}{E} \right]$. That sum gives us χ^2, our test statistic:

$$\frac{(7 - 5)^2}{5} + \frac{(70 - 75)^2}{75} + \frac{(23 - 20)^2}{20} = 1.583$$

After χ^2 has been calculated, we look up the critical value on the proper table. This is a table derived from χ^2 distributions. Like the t-distributions, there are many. In the case of the t's, we had one distribution for each sample size. In the case of the χ^2's, we have one distribution for each number of classes (or **cells,** as they are called in this case) in the distribution. It is therefore essential that the observed and expected distributions have the same number and size of classes or cells. The χ^2 distribution, then, is a distribution of the differences between distributions that can be expected at different levels of probability. To use the table, we need to know P. We also need to know something called degrees of freedom. This is a term that is used for many hypothesis testing situations. We will only say here that in this use of the chi-square test, the **degrees of freedom** (abbreviated df) is always equal to one less than the number of classes or cells. In our current example, there are 3 cells, so we have 2 degrees of freedom. Since our significance level is .05, we look under a = 05 and df = 2 to find the critical value of 5.99147. Since our test statistic is much smaller than that, we know the difference between the two distributions is not significant so we accept our null hypothesis.

We are now ready to list the steps in a χ^2 test and then to look at another example.

Steps for a Chi-Square Test

1. State the null and alternative hypotheses.
2. Decide on a significance level.
3. Look up the critical value on the table, using the proper significance level and degrees of freedom.
4. Calculate the test statistic.
5. Compare the test statistic and the critical value. If the test statistic is larger, reject the null hypothesis. If it is smaller, accept it.

Example 4: Suppose we take a random sample of 200 families in a study of consumer spending patterns. We want to make sure that the income distribution of our sample matches that of the population. The income figures are as follows:

	Number in Sample	Percent of Population
Under $5,000	16	12%
$5,000–$14,999	80	44%
$15,000–$24,999	65	30%
$25,000 and over	39	14%

1. **Hypotheses.** H_o: There is no difference between the income distribution of the sample and the population.

   H_a: There is a difference between the income distribution of the sample and the population.

2. **Significance Level.** Set P = .05

3. **Critical Value.** Since χ^2 is a measure of difference that does not distinguish between larger or smaller than expected, all tests are one-tailed. Therefore the area (a) is the same as the significance level (P) and we use .05. Because there are 4 cells, df = 4 − 1 = 3. The value from the table (in text, p. 307) therefore is 7.81473.

4. Test Statistic.
$$\chi^2 = \Sigma \frac{(O - E)^2}{E}$$

Notice that since the population distribution is in terms of relative frequencies, we must translate this into actual expected frequencies for a sample of 200. So 12% of 200 = 24, 44% of 200 = 88, 30% of 200 = 60 and 14% of 200 = 28.

$$\chi^2 = \frac{(16 - 24)^2}{24} + \frac{(80 - 88)^2}{88} + \frac{(65 - 60)^2}{60}$$

$$+ \frac{(39 - 28)^2}{28} = \frac{64}{24} + \frac{64}{88} + \frac{25}{60}$$

$$+ \frac{121}{28} = 8.13203$$

5. Conclusion.
Since the test statistic is larger than the critical value we reject the null hypothesis and accept the alternative. That is, there is a difference between the income distribution of the sample and the population.

The chi-square test can also be used to compare two samples. In that case, one of the samples becomes the source of observed and the other of expected data. In the exercise below, assume that we are comparing the gadget makers to the widget makers. The widget makers will then be the source of expected data.

Exercise: The table below shows the age distribution among a sample of 100 workers at a widget factory and a sample of 100 workers at a gadget factory. Use a chi-square test to see if the two distributions differ significantly.

Age	18–26	27–40	over 40
Population	52%	39%	9%
Sample	46	43	11

PART B: Hypothesis Testing Headaches

Although the three examples of hypothesis testing that we have just discussed are fairly clear cut, there are a number of problems, both theoretical and practical, that arise when these techniques are applied in real situations. In Chapter 1, we looked at a number of inductive fallacies in a fairly general way. Now we can look at them again in the specific context of hypothesis testing.

Status of the Conclusion. We stated in Chapter 1 that statistics is an important tool for inductive reasoning. We can now see why this is the case. The statistical technique of hypothesis testing allows us to look at the conclusion of an inductive argument (where the conclusion is that we either accept or reject the null hypothesis) and calculate the probability of that conclusion being correct. Remember that in the case of inductive arguments, that is, arguments that depend on collecting evidence and drawing conclusions from that evidence, we can never prove anything conclusively. All we can do is build up a case by collecting more and more evidence. Now hypothesis testing can give us a more precise idea of how much faith we can place in a conclusion by assigning a probability to its being correct. We must not let this seeming precision lull us into a state of complacency. Remember that the probability refers only to the probability of the null hypothesis being correct. We must look now at the relationship between the null hypothesis and the whole argument.

Assumptions. Often the null hypothesis of a research project is connected with the real question being asked by a string of assumptions which may or may not be explicitly stated. Sometimes these assumptions are questionable at best. Look back at Example 21 in Chapter 1. This example concerned an ad for a brand of scotch which was claimed to be better because it was 90 proof. The argument was that after a given period of time during which it was diluted with melting ice it would still have a higher proof than an 80 proof scotch and therefore taste "fresher." The hypothesis test would probably give precise results. There would indeed be a significant difference between the two. But what about the assumption that higher proof determines the subjective experience of "freshness" — not to mention the question of why the particular scotch being advertised should be better than any other 90 proof. Consider the following example:

Example 5:   In *The Psychology of Sex Differences,* Eleanor E. Maccoby and Carol N. Jacklin review a variety of psychological studies of sex differences. One set of experiments seeks to answer the question, "Do girls show more dependence than boys?"

> Two reports of studies that were carried out in field situations by anthropologists . . . show records sampling how far away from home children aged 3–8 were found during their free time, when they had a choice of where to play. In two different cultures in Kenya, boys customarily played at a greater distance from home.

The hypothesis testing situation is quite clear. The average distance for boys can be calculated and compared to the average distance for girls. We can then see what the probability is that there is a real difference. However, what does this mean about dependence? As Maccoby & Jacklin point out,

> It is difficult to know how to interpret these findings. . . . Do girls stay closer to home because

they want to be able to run to the mother for security in case any threatening situation arises? Have they been given more warnings than boys about the dangers of the outdoor environment? Are the kind of settings they are allowed or expected to participate in located closer to home? We do not know. (pg. 196–197)

It is very important then, to look not only at the actual null hypothesis being tested, but to see what the relationship is between that hypothesis and the larger question being investigated.

Suppressed Evidence. Another example of fallacious reasoning we discussed earlier was a selection from all information available that led to a distortion or even a contradiction of the real situation. In Chapter 1 we looked at some examples where this technique was used intentionally to make an argument more convincing. It can also happen as a result of oversight or concentrating on one aspect of a problem. An interesting example is the following.

Example 6:   Thomas McKeown, writing in the April 1978 issue of *Human Nature,* discusses some of the factors influencing the declining death rates in this century. His point is that we tend to think of modern drugs as having made most of the difference. As an example, he looks at the decrease in deaths from tuberculosis after the discovery of streptomycin. He states that "streptomycin (chemotherapy) lowered the death rate from tuberculosis in England and Wales by about 50 percent." So if we did a hypothesis test comparing the death rates before and after the use of this drug, we would get a highly significant difference. However, he goes on to point out that "its (the drug's) contribution to the decrease in the death rate since the early 19th century was only about 3 percent." (pg. 63)

The graph on the next page shows the complete picture.

We need not belabor the point that the result of testing a null hypothesis is only as valuable as the reasoning that went into the design and execution of the experiment. As the computer programmers so succinctly put it, "Garbage in, garbage out." If the data we are using for the test is basically irrelevant to the question we are investigating, we are wasting our time. If the sample from which we get the data is not a truly random sample, the results will be affected. If we try to overgeneralize, we will be in trouble, and so on. But besides these problems which are common to all inductive arguments, the hypothesis testing technique has some problems peculiar to itself that we should look at.

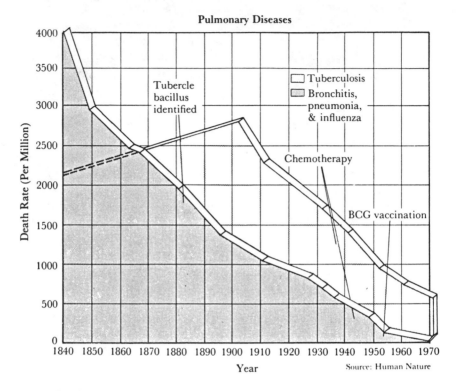

Pulmonary Diseases

Source: Human Nature

The Appropriate Test. In this chapter we have looked at three kinds of hypothesis tests: a one-sample means test for large samples, the same for small samples, and the chi-square. These are only a few of the hundreds of tests that statisticians use in as many different situations. It is beyond the scope of this book to explore more tests in detail, but it may be useful to get an idea of what it is that determines which test is used.

Statistical tests can be divided into two main categories — parametric and non-parametric tests. **Parametric tests** are based on the use of population parameters, such as the mean, variance or standard deviation. The data consist of continuous measures, such as test scores, reaction times, distances, etc. A necessary assumption in most cases is that the variable being measured is normally distributed. The one sample means test and the t-tests discussed above are examples of parametric tests. **Non-parametric tests** are used when we are dealing with categorical data, where we count the number of people in different categories (Democratic, Republican, rich, poor, etc.) or when we are dealing with a variable which is not normally distributed. The chi-square test is an example of a non-parametric test. If a parametric test is used on categorical data or with a highly skewed distribution, the results are invalid. If a non-parametric test is used on normally distributed continuous data, the results are valid but less powerful than if a parametric test had been used.

A second way in which tests are classified is according to the source of the data. In the examples discussed here, we compared, in each case, a sample to a population. In actual practice, it is more common to compare two samples. When this is done, appropriate adjustments must be made in the theoretical distribution used to calculate the critical value. Furthermore, the relationships between the two samples must be considered. Are they independent random samples, or were they matched according to some characteristic? Were they gathered from two different groups or from one group on different occasions? Each of these situations calls for a different test.

Hypothesis testing, when used correctly, can give us important information about the probability of a carefully stated hypothesis being true. However, in using this technique, we must be careful that the conditions of the test are met, that the data is both appropriate and accurate, and that we do not make too great a leap from the particular information given us by the results of the test to very general questions about the nature of the world.

SUMMARY

PART A

A **hypothesis** is a statement which can be tested to see if it is true or not. A **null hypothesis** is a hypothesis which is usually stated in the form: "There is no difference between X and Y" where X and Y are numerical measures obtained from samples or from samples and populations. The **alternative hypothesis** is the statement that there is a difference. In hypothesis testing situations, the null hypothesis is tested to see if it should be accepted or not. If it is rejected, then the alternative hypothesis is accepted. The **test statistic,** which is a number calculated from the sample, is used to decide whether or not to accept the null hypothesis. The **critical value** is the value that the test statistic is compared to in reaching this decision. The **significance level** is the probability of getting the critical value by chance, and is used to calculate the critical value, since it is set in advance. The critical value is calculated from the proper sampling distribution. The area of the graph determined by the critical value, which tells us to reject the null hypothesis if the test statistic falls within it, is called the **rejection region.** If there are two rejection regions, one at each end of the graph, the test is called a **two-tailed test.** If there is only one rejection region it is a **one-tailed test.** There are two types of error which can be made: The **Type I** error of rejecting a true null hypothesis and the **Type II** error of accepting a false null hypothesis. The smaller the chance of making a **Type I** error, the larger the chance of making a **Type II** error and vice versa. The probability of making a **Type I** error is equal to the significance level.

A **one-sample means test** is a test which compares the mean of a sample to the population mean to see if there is a real difference. When the sample used in any hypothesis test is of size 30 or less and/or σ is unknown, test using a **t-distribution** rather than the normal distribution. There is a different t-distribution for every sample size. A **chi-square test** is used to compare the frequencies of two distributions to see if they are different.

PART B

There are a number of ways in which the technique of hypothesis testing can be misused. First of all, it is important to understand that the null hypothesis is never proved nor disproved, it is accepted or rejected, with a certain probability of error. Also it is important to realize that the probability refers to the truth of the null hypothesis, not to the larger general question to which the null hypothesis may be connected by a string of assumptions. The conclusion that is reached is only as valid as the entire argument that supports it, and all the possible defects associated with inductive arguments, as well as those associated with sampling techniques, must be carefully avoided.

It is also important that the proper statistical test be employed. **Parametric tests** can be used only with continuous measures based on population parameters such as mean, variance or standard deviation, and these must be normally distributed. Otherwise **non-parametric tests** must be used.

EXERCISES

PART A

1. *The mean effective tax rate in Cambridge is about 5% of the market value of real estate, with a standard deviation of 1%. If a landlady who owns 144 properties in the town is found to pay a mean effective tax rate of 4.5% on her properties, is she getting an unfair advantage? Use the one-sample mean test to decide.

2. Another landlady in Cambridge is also paying a mean of 4.5% tax, but she owns only 9 properties. Use a t-test to see if there is sufficient evidence to accuse her of having an unfair advantage.

3. *Calculate the significance level for the test statistics in the above two examples. What can you conclude from these?

4. Below is a table showing the age distribution of the freshman class of a

large university and the age distribution of a sample of 50 students taken from that class. Use a chi-square test to decide if the sample distribution is close enough to the population distribution to be considered representative.

Age	18	19	20	over 20
Population	40%	27%	10%	23%
Sample	21	14	10	5

5. *A new sample is taken in the survey above and now there are 21 eighteen year olds, 15 nineteen year olds, 4 twenty year olds and 10 older students. How well does this group represent the population?

6. What is the minimum amount of information necessary to do a one sample means test? To do a chi-square test?

7. Devise an example of your own for a t-test and a chi-square test.

PART B

1. *Consider the situation in Example 1 in this chapter. Assume that $\bar{x} = 1.55$ while $\mu = 1.5$ and $\sigma = .5$. The doctor performs a hypothesis test and finds that the number of days absent with respiratory disease is no different for asbestos factory workers than for anyone else. She therefore concludes that working in the factory is not injurious to the workers' health. What is wrong with this argument?

2. Below are several different situations. In each case, see if the information given is appropriate for one of the three tests described in this chapter. If so, perform the appropriate test. If not, say why not.

a. The following table appeared in the May/June 1978 issue of *Working Papers* (pg. 60). Is there a significant difference in the unemployment rates for the U.S. and Canada?

EFFECTS OF THE 1975 RECESSION
AN INTERNATIONAL COMPARISON

	AVERAGE ANNUAL GROWTH RATES (%) (real per-capita domestic output, 1970 prices)		UNEMPLOYMENT RATES (%)			
	1966–73	1973–75	1973	1974	1975	1976
United States	2.7	−2.7	4.9	5.6	8.5	7.7
Canada	3.5	.4	5.6	5.4	6.9	7.2*
U.K.	2.5	− .5	2.3	2.4	3.8	5.3
France	4.9	0.	2.0	2.3	3.9	4.3
Germany	3.9	−1.5	1.0	2.2	4.1	4.1
Japan	9.1	− .6	1.3	1.4	1.9	2.0*
Australia	—	—	1.9	2.2	4.4	5.2*

* Third quarter, 1976

Source: Data published by the Organization for Economic Cooperation and Development in *National Accounts* for 1962 to 1975, *Statistical Telegram,* July 1977, and *Quarterly Supplement,* November 1976.

b.   The following report on food costs appeared in the *Boston Globe* of June 28, 1978 (pg. 58). Is there a significant difference between the market basket price for Tampa and Des Moines?

17-City Market Basket Comparison (June 1)			
City	Food Cost	Tax (if any)	Total	difference from average
Tampa	$32.58	0–0	$32.58	−12.0%
Des Moines......................	$33.80	0–0	$33.80	−8.5%
San Diego........................	$34.02	0–0	$34.02	−8.0%
Phoenix	$33.19	5%–1.66	$34.85	−5.7%
Atlanta	$34.60	4%–1.38	$35.98	−2.7%
Cleveland	$36.08	0–0	$36.08	−2.4%
Dallas	$36.41	0–0	$36.41	−1.5%
New York	$37.72	0–0	$37.72	+2.1%
Portland, Ore.	$38.10	0–0	$38.10	+3.1%
Chicago	$36.47	5%–1.82	$38.29	+3.4%
Little Rock	$37.36	3%–1.12	$38.48	+4.1%
San Francisco..................	$38.82	0–0	$38.82	+5.0%
Philadelphia....................	$38.88	0–0	$38.88	+5.2%
Salt Lake City	$37.12	5%–1.86	$38.98	+5.5%
Washington, D.C.	$38.99	0–0	$38.99	+5.9%
Boston.............................	$39.40	0–0	$39.40	+6.6%
Anchorage	$50.21	0–0	$50.21	+35.9%

c.   The number of traffic accidents in a certain town decreased from 27 per year to 23 when a traffic light was installed. Did the traffic light make a significant difference?

d.   A sample of 25 people had a mean income of $15,000. If the population mean is $13,700 with a standard deviation of $5,000, is the sample mean significantly different from the population mean?

Chapter 12

ESTIMATION

Statistics, as we have said before, provides techniques for getting answers about populations by looking at samples selected from those populations. In the last chapter we saw that hypothesis testing was one way to do that. We found, however, that hypothesis testing required our having quite a bit of information to begin with. In tests involving the mean, for instance, we needed to know at least one population parameter, the mean, and preferably the standard deviation as well. In many real life situations information about the population is simply not available. It is then that the other major type of statistical inference is used — estimation. Estimation procedures allow us to use information gained from a sample to make inferences about the population. Any given population parameter, such as the mean, range, standard deviation or binomial probability can be estimated. We will look at the techniques for estimating two important parameters, the mean and the binomial probability.

PART A: Estimating Population Parameters

Estimating the Mean

Suppose we wanted to know the average age of the children in a large day care center. How would we go about estimating that mean? We might ask a group of nearby children how old they were and get a sense of their average age. If we happened to know the ages of the oldest and the youngest child in the center, we could assume the average age is half way in between. Or we could take a random sample and find its mean, median or mode. Each of these methods might be said to give an estimate of the population mean, but some would surely be better than others. How do we choose?

Estimators. Any procedure or formula for finding an estimate of a population parameter is called an **estimator.** To see which of the possible estimators to use for a given population parameter, mathematicians use a number of different criteria to see which is the best estimator. One such criterion is that the estimator must be unbiased. An **unbiased estimator** is one which has a sampling distribution whose mean is the population parameter. So for example, an unbiased estimator for the population mean is the sample mean. This is true because the mean of the sampling distribution of the mean is in fact always the population mean. Another criterion is efficiency. An **efficient estimator** is one whose sampling distribution has the smallest possible standard error. Again the sample mean meets this criterion. We therefore say that the sample mean:

$$\bar{x} = \frac{\Sigma\, x}{n}$$

is the best estimator of the population mean, μ.

In the case of the mean, then, we estimate the population parameter by applying the same formula to the sample data that we would use with the population data if that were available to us. However, we do not always use the formula for the population parameter and apply it to the sample. An important exception is the estimator for the population standard deviation. The formula for the population is:

$$\sigma = \sqrt{\frac{\Sigma\, (x - \mu)^2}{n}}$$

But as we have seen in Chapter 7, when we want to estimate the standard deviation of a population from sample data, we use the formula:

$$s = \sqrt{\frac{\Sigma\, (x - \bar{x})^2}{n - 1}}$$

Of course it is not necessary for the student of statistics to make his own decision about the best estimator for any particular parameter — that work has already been done and the details need not concern us here. However, once we know which estimator to use, we must still concern ourselves with determining how accurate an estimate we can expect to get from it in any particular case.

Bound on Error of Point Estimates. We know that the best estimator for the mean of a population is the mean of a random sample of that population. Now we want to see how accurate we can expect such an estimate to be. Let's look at an example.

Example 1:　Assume that we want to estimate the mean age of all children in nursery school in our city. We take a random sample of 100 nursery school children and calculate the mean of that sample. We find that it is 3.75 years. How accurate is that estimate?

In order to judge the accuracy of an estimate, we must turn again to the sampling distribution. In this case, we have a sample mean (3.75 years) which we are using as our estimate of the true mean of the whole population. How close is the sample mean likely to be to the true mean? We know that in any sampling distribution of the mean, there is a probability of .95 that any given sample will fall within two standard deviations of the population mean. What we need to know, then, is the size of the standard deviation (standard error) for this particular sampling distribution.

In the past we have used the formula

$$s.e. = \frac{\sigma}{\sqrt{n}}$$

However now we are in a position where we don't know σ, the standard deviation of the population. We can, however, estimate it with s, the

standard deviation of the sample, where $s = \sqrt{\dfrac{\Sigma (x - \bar{x})^2}{n - 1}}$

Assume that we perform the calculation on our sample and find that

$s = .8$. Then $s.e. = \dfrac{.8}{\sqrt{100}} = .08$.

Since 95% of the time the sample mean is within two standard deviations of the population mean, we have a 95% probability that the population mean is between $3.75 + 2(.08)$ and $3.75 - 2(.08)$. So $2(.08)$ or .16 years is the amount of error we can expect in either direction. We can then state that the estimated mean age of the children in day care centers is 3.75 years with a bound on the error of estimation of .16 years.

Exercise: Suppose we want to estimate the mean age of adults (18–65) who voted in the last presidential election in Philadelphia. From a random sample of 100 voting adults we found a mean of 37.5 and a standard deviation of 11. Find the bound on the error of estimation assuming a 95% probability level.

Confidence Intervals. What we've just done is to estimate a certain point for our population parameter (in this case the mean age of nursery school children), and to say, "Here's our estimate, and here's how close to reality we can be pretty sure that it is." By analogy, this is like saying, "I'll pick you up at 2:00 pm, but I might easily be as much as ten minutes early or late." A confidence interval is another way of stating the same estimate. Instead of defining a point and a bound on error, we define an interval within which we expect the population parameter to lie. This is analogous to saying, "I'm pretty sure I can pick you up between 1:50 and 2:10 pm."

Confidence intervals are defined in terms of percentage probabilities. If we have a probability of .95 that the population parameter falls within a certain interval, we call this a **95% confidence interval.**

Example 2: Using the same sample of 100 nursery school children as in Example 1, what is the 95% confidence interval for the population mean?

We use the same calculations as we used in Example 1, which told us that we're 95% sure the population mean is no more than .16 away from 3.75. So our 95% confidence interval is 3.75 ± .16, which is to say the interval between 3.59 and 3.91. The standard form for writing a confidence interval is like this: [3.59, 3.91].

There is nothing sacred about the 95%. It is simply the most commonly used confidence interval, just as 5% is the most commonly used significance level. Actually, a 5% significance level and a 95% confidence interval are just two different ways of saying the same thing. If there is a 5% chance of the sample mean falling *more* than two standard deviations from the population mean, there is a 95% chance of it falling *less* than two standard deviations away. While the significance level is the chance of error, the confidence level is the chance of being correct.

We could, and sometimes do, use a 90% or a 99% interval. But notice that there is a tradeoff here between how much confidence we can have in capturing the true mean in our interval, and how useful our estimate is. We might insist on being 99.9% sure that the population mean was somewhere within our interval, but our interval might have to be so large that it didn't really tell us much about the population we are trying to study. A commonly used analogy is that of a marksman trying to hit a target. The chances of hitting a very large target are much better than those of hitting a small one. But hitting a large target doesn't prove much.

Exercise: Using the same data as in the previous exercise, find the 95% confidence interval for the population mean.

Small Sample Estimates

Next we want to look at estimates involving small samples. As we have said before, the assumptions of a normal distribution cannot be expected to hold if our sample size is less than about 30. Therefore in estimation, as in hypothesis testing, we make use of the t-distribution for small samples. Let's see how we can estimate a mean using this method.

Example 3: Suppose we are again looking at the average age of children in a day care center but this time we are only able to take a sample of 16 children. We calculate the mean and standard deviation of the sample and find them to be $\bar{x} = 3.75$ and $s = .8$. We again want a 95% confidence interval.

With large sample estimation we use a normal distribution so we know that 95% of our sample means will lie between the points two standard deviations to the left and two standard deviations to the right of the population mean. In the small sample case, we cannot use that assumption, so we must find the t-scores of the points which enclose 95% of the sample means. We do that by consulting the table of t-distributions. We need to know a and n to use the table. Since a 95% confidence interval corresponds to a rejection region of 5% distributed between the two tails of the graph, we want an area of half of .05, or $a = .025$. Looking down that column to the point where $n = 16$, we find that $t = 2.131$. In other words, 95% of the sample means will fall within 2.131 standard deviations to the left or right of the population mean. Therefore we can calculate the 95% confidence interval as: 3.75 ± 2.131 $(s.e.)$ Since $(s.e.) = s/\sqrt{n} = .8/\sqrt{16} = .8/4 = .2$, we get: $3.75 \pm 2.131(.2) = 3.75 \pm .43$, or the interval [4.18, 3.32]. We can also express this result as the point estimate and the bound on error. Then we say that the point estimate of the population mean is 3.75 and that the bound on error is .43.

Exercise: Now suppose we had taken another random sample with 25 children. We found the mean to be 4.00 and the standard deviation equal to 1.1. Find the 95% confidence interval for the population mean.

The steps for estimating any population parameter are basically the same.

Estimating Population Parameters

1. Calculate the estimator for the sample.
2. Decide on the probability for the confidence interval or probability for the bound on the error of estimation.
3. Use the sampling distribution of the estimator to calculate the actual bound on error or confidence interval for a large sample, the appropriate t-distribution for a small sample.

Estimating the Binomial Probability

As another example of a parameter we might want to estimate, let's look at the binomial parameter, p. Recall from chapter 10 that p represents the probability of a given outcome for an event when there are only two possible outcomes. The sampling distribution for p is given in Figure 1.

Figure 1

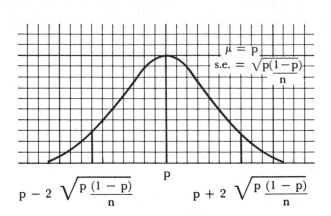

$$p - 2 \sqrt{\frac{p\,(1-p)}{n}} \qquad p \qquad p + 2 \sqrt{\frac{p\,(1-p)}{n}}$$

Recall that p is the probability of the desired event. To find the standard deviation, we multiply the probability that the desired outcome will occur by the probability that it won't, divide by the sample size and then take the square root. Using these characteristics of the sampling distribution, we can now estimate p.

Example 4: A public opinion poll sampling 121 voters found that 77% of them were dissatisfied with their local government officials. How accurately does this poll reflect the population's feelings?

We use $p = .77$ as our estimate.

Then we must calculate the standard deviation of the sampling distribution:

$$s.e. = \sqrt{\frac{p(1-p)}{n}} = \sqrt{\frac{.77(.23)}{121}} = .038$$

So the 95% confidence interval is $.77 \pm 2(.038)$ or approximately the interval [.69, .85]. Another way we could say it is that about 77% of the population is dissatisfied, with a bound on the error of that estimate of about 8%.

PART B: Estimation Errors

Since so many of the statistics that rule our lives — unemployment figures, economic indicators, public opinion polls — are the result of estimation procedures, it is important to recognize the possible shortcomings of published estimates.

Sample Data. As in the case of hypothesis testing, the results of any estimation procedure can be only as good as the data it is based on. Advertisements are traditionally suspect, but even sources we are least likely to doubt can be guilty of mammoth errors. The following is by no means an isolated incident, and articles like this can be found with alarming frequency in the larger newspapers.

Example 5: *The New York Times* of July 29, 1974 reported the following story under the headline, "Data Revisions Dismay Experts:"

> Economists in and out of the Government are expressing concern, even dismay, at the huge size of the revisions in the nation's basic accounts published by the Commerce Department earlier this month.
>
> "It makes you feel like a fool," said one high-level Government economist the other day. "Our analyses and forecasts were based on numbers that were badly wrong." (pg. 33)

The largest error in this case was in the estimation of inventories—the value of goods produced but not sold. The original estimate of $5.5 billion had to be revised to $16.9 billion!

Suppose, however, that the data on which the estimate is based are as good as can be obtained. There remains the problem of the bound on the error of estimation, something almost never mentioned when the estimates are reported.

Size of Error. One of the areas in which the size of the expected error can be important is that of opinion polls. The large national polls like Gallup and Harris are usually based on a sample of about 1500 people, but many local polls are based on much smaller samples. And we know that the size of the error is critically related to the size of the sample. Yet we rarely see, in the news stories which report these polls, what the bound on the error of estimation is, although (or maybe because) it is often of considerable size.

The size of the error of estimation is particularly important when a comparison is being made. Suppose we are looking at the results of two polls, taken at different times. If we want to say that the situation has changed from one time to the next, then we must make sure that the difference between the two polls is not just the result of sampling error. We know that if we take two samples from a population there is likely to be some difference between them, but that 95% of the samples will give results within two standard deviations of the mean. So if the difference between the results of the two polls is less than two standard deviations, we have no way of knowing whether there is a real difference or just the normal fluctuation from one sample to the next. In order to be confident that we have a real difference, the size of the difference must be bigger than the bound on the error of estimation. This is often not the case, even when we are dealing with large national polls.

Example 6: On August 12, 1974 *The New York Times* published the following graph of a large number of polls that were taken to measure President Nixon's popularity over a period of time. During 1973–74, for example, 33 different polls were taken. Each of these polls was reported in the papers with a story which attempted to explain the most recent rise or fall in the president's popularity. But how many of these fluctuations were really significant? Let's look at the last ten polls that are graphed, beginning with "Ford Confirmed."

Figure 2

Nixon's Popularity as Measured by the Gallup Poll

Question: Do you approve of the way Nixon is handling his job as President?

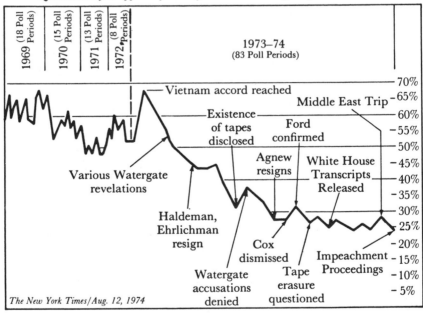

The New York Times/Aug. 12, 1974

What we have, statistically speaking, is a series of binomial samples in which the two categories are approving or not approving. Approving is the desired event. We can figure the bound on the error of estimation quite easily. It is safe to assume that samples were at most of size $n = 1500$, the usual size for Gallup polls. The percentages favoring the President hover around 27% in the last ten polls, so we can take the estimate for p to be .27, which makes $1 - p = .73$. The standard deviation of the sampling distribution, then is:

$$s.e. = \sqrt{\frac{.27(.73)}{1500}} = .0115$$

The bound on the error of estimation at a 95% probability level, is $2(.0115) = .023$. That means that the estimates are accurate to within 2.3% either way. We see that all fluctuations except the first and last of the series are *smaller* than the bound on the error. So we have no way of knowing whether these fluctuations reflect real differences in public opinion or are simply chance fluctuations from one sample to the next. Perhaps other, equally random samples would have produced quite different fluctuations, and the newspapers would have discovered different events which equally plausibly explained them.

Point vs. Interval Estimates. We see then that point estimates are sometimes misleading. They allow us to believe that an estimate is much more precise than it actually is. Of course no one would want to look at a poll and see interval estimates of the percents: "Between 25% and 30% approve the President's recent decision" would make clumsy reading. And for most purposes the greater accuracy of that statement would not make enough difference. In other cases it is hard to understand why intervals are not used.

Example 7: The following is quoted from an article on EPA tests of gasoline mileage ratings of new cars. It appeared in the *Boston Globe,* September 20, 1977:

> He [Mr. Costle] said nearly 50 percent of motorists get mileages as good or better than the EPA tests but "there are a large number of drivers who do not get the advertised mileage."
>
> This is due to "wide differences in driving habits and maintenance practices and the varying road, traffic and weather conditions which drivers experience," he said. In testing 1979 model cars, EPA will adjust the figures downward, possibly taking off a certain percentage to account for variables.
>
> Costle also suggested consumers should use the EPA mileage figures not as a promise of performance but as a guide for comparison shopping — to indicate, for instance, that one car might use 20 percent more fuel than another.

There are two problems here. The consumer would like to be able to compare different cars to see which gets the best mileage. Once the selection is made, it would also be helpful to know what actual mileage can be expected. At present neither of these goals is being met very adequately, and reducing the average by a set percentage certainly wouldn't do it either. If it is true that 50% of motorists actually get mileages as good as or better than the EPA tests, then the estimates given are probably as accurate as they can be already. What is needed is to give the consumer a better sense of the range of possible values for the mileage. Here interval estimates would really be helpful. Instead of expecting the new car to get exactly 32 miles per gallon on the highway, the consumer would have the more realistic expectation of getting, say, between 30 and 34 miles to the gallon. Similarly in comparison shopping, interval estimates would be less likely to mislead the consumer into thinking that there is a real difference between a car that was estimated to get 33 miles to the gallon and another that was estimated to get 34. Not enough is known about testing practices by the general public to make such judgments possible. Interval estimates would make the true state of affairs much clearer.

Finally, it is important to remember that for a given sample, any increase in the chance of being right necessarily means a more vague estimate and vice versa. The only way to get an estimate that is at the same time both more precise and more probably accurate is to increase the size of the sample. However, this is expensive, and the added cost must be weighed against the potential benefit to be gained by increasing the accuracy of the estimate.

Estimation is one of the most useful statistical procedures. But as with any statistical inference, we must be careful not to accept any conclusions uncritically. An estimate is only as good as the data that went into it and only as accurate as the bound on its error.

SUMMARY

PART A

Estimation is a procedure which uses information from a sample to approximate a population parameter such as the mean, standard deviation or binomial probability. An **estimator** is a rule or formula for calculating the estimate of a parameter. An **unbiased estimator** is one which has a sampling distribution whose mean is the population parameter. An **efficient estimator** is one whose sampling distribution has the smallest possible standard error. The best estimator for the mean is $\bar{x} = \dfrac{\Sigma\, x}{n}$. For the standard deviation, it is $s = \sqrt{\dfrac{\Sigma\,(x - \bar{x})^2}{n - 1}}$. The accuracy of an estimate is indicated by the bound on the error of estimation. Since the sampling distribution tells us that 95% of all sample means fall within two standard deviations of the population mean, in a large sample estimate, the bound on the error of estimation is taken to be $2(\frac{s}{\sqrt{n}})$. That is, we use the sample mean as the estimate of the population mean and expect that (with a 95% probability) the true value will lie no farther than two standard deviations in either direction. If other probability levels are used, the bound on error will change proportionately. An interval estimate is formed when we add and subtract the bound on error from the mean. The **95% confidence interval** for the mean is $\bar{x} \pm 2(\frac{s}{\sqrt{n}})$ if the sample is larger than 30. For smaller samples we use the t-distribution. Then the bound on error is $t(\frac{s}{\sqrt{n}})$ and the 95% confidence interval is $\bar{x} \pm t(\frac{s}{\sqrt{n}})$. In the case of a binomial distribution, the estimator for the population parameter p is the value of p derived from the sample with a bound on the error of estimation of $2\sqrt{\frac{p(1-p)}{n}}$ at a 95% probability level for a large sample, and $t\sqrt{\frac{p(1-p)}{n}}$ for a small sample. The 95% confidence interval is $p \pm 2\sqrt{\frac{p(1-p)}{n}}$ for a large sample and $p \pm t\sqrt{\frac{p(1-p)}{n}}$ for a small sample.

PART B

In estimation, as in hypothesis testing, the results are only as good as the data on which they are based. The size of the error is important, especially if a comparison is being made between the same measure taken at different times or from different populations. If the difference between measures is not larger than the bound on the error of estimation, then no real difference can be inferred to exist. In some cases, using the interval estimation procedure will give a truer picture than using a point estimate, while in other cases a point estimate will be more convenient.

EXERCISES

PART A

1.a. *Estimate the mean price of a loaf of rye bread in a city where a survey of 49 stores shows that the mean is 72¢ with a standard deviation of 3¢.

  b. What is the bound on the error of estimation at a 95% probability level?

  c. What would the 90% confidence interval for the price of rye bread be?

2.a. Suppose that only 9 stores were sampled in the above survey. Then what would the estimated mean price be?

  b. What would be the bound on the error of estimation at a 99.7% probability level? ($n = 9$)

  c. What would the 95% confidence interval be? ($n = 9$)

3. *A survey of 400 consumers shows that 89% are dissatisfied with the quality of the new cars they have just bought.

  a. What would the estimated probability for the population be?

  b. What is the bound on the error of estimation? (95% probability level)

  c. Find the 95% confidence interval for p.

4. Suppose you wanted to estimate the mean number of children per family in your town. Explain in detail what steps you would take to find that estimate and how you would express the results.

5. *If you were the campaign director of a local candidate for office, how would you estimate her chances of being elected. Explain in detail.

PART B

1. What is some of the information you would want to have before accepting the following estimates?

  a. *From the *Boston Globe,* July 19, 1978:

    "Americans ate 43.3 pounds of chicken per person in 1976 compared to 15.5 pounds in 1910." (pg. 65)

  b. From the *New York Times,* September 8, 1977:

    "Last week, the Labor Department reported that 14.5 percent of all black workers were unemployed in August, which equaled a postwar high." (pg. A17)

  c. *From the *Boston Globe,* July 20, 1978:

    "About 19 percent of those under 25 were dissatisfied with their jobs compared with 7 percent of those over 55." (pg. 31)

2. What is wrong with the following argument:

  Candidate Smith is gaining ground against candidate Jones. This month a poll of voters showed that 47% favor him as opposed to only 45% last month. In each case a random sample of 100 people was used.

3. Give two examples where point estimates would be more appropriate and two examples where confidence intervals would be better. Explain why.

Chapter 13

LINEAR REGRESSION

The kinds of statistical inferences discussed so far have all dealt with one measurement at a time. We have looked at the mean number of days missed from work or the average cost of a loaf of bread, and have estimated the age of children or the probability of a candidate winning an election. But much scientific work is concerned with the relationship between two sets of measurements. A doctor might want to know whether there was a relationship between the amount of exposure to pollutants and the number of days missed from work. Or an economist might investigate the relationship between the amount of wheat planted and the cost of a loaf of bread. These last two chapters will deal with the problem of relationships between two sets of data.

When we are concerned with relationships, there are several questions we are interested in answering. First of all, we want to be able to describe the relationship. If it is between two sets of measures, we will want to know how they are related. Does an increase in one measure mean an increase or a decrease in the other? By how much? Can we use one set of data to predict or estimate the other? How accurately will it do that? There are two statistical concepts we use in trying to answer these questions.

Linear regression techniques help us to describe the relationship between two sets of measures and to estimate one set from the other. Then, if we want to know how accurate that estimate is, or how significant the relationship between the two measures really is, we use the concept of correlation. Linear regression will be taken up in this chapter and correlation in the next.

PART A: Finding the Regression Line

Scatter Diagrams. The best way to see the relationship between two sets of measures is to use a graph. In Chapter 3 we saw that the simplest and least dramatic graph possible was the point graph. This type of graph turns out to be the most useful way of representing the relationship between two sets of data for the kind of analysis we will want to do. Remember that a point graph allowed us to represent two pieces of information with each point, one piece of which could be read off the horizontal scale and the other off the vertical. The point graphs we encountered in Chapter 3 tended to have no more than one point above any place on the horizontal axis. This restriction does not hold for the graphs we will be using here. Another difference is that while the point graphs we encountered before used only positive numbers, we will now expand our graph to include negative values where needed. Numbers to the left of the vertical scale and below the horizontal scale will be negative, while the positive scale will be as before. These graphs, used to record the relationship between two sets of measures, are called scatter diagrams. Let's look at an example.

Example 1:   A sociologist is interested in seeing how well women's perceptions of legislators correspond to the legislators' actual performance. She chooses sixteen legislators and records how many favorable votes each cast on a total of ten bills supporting women's rights. Then she asks a group of women to rate the same sixteen legislators on a scale of 1 to 10, where a score of one indicates no sensitivity to women's issues and a score of 10 indicates high sensitivity. The data looks like this:

Legislator:	A	B	C	D	E	F	G	H	I	J	K	L	M	N	O	P
Positive Votes:	1	1	2	2	3	3	4	4	5	5	6	6	7	7	8	8
Mean Score:	1	2	1	3	2	4	3	4	4	5	6	7	8	9	10	7

We make a scatter diagram by marking off the horizontal axis in units for the number of votes and the vertical axis for scores. Then we plot a point for each pair of numbers — one point for each legislator: (1,1) (1,2) (2,1) (2,3) (3,2) (3,4) (4,3) (4,4) (5,4) (5,5) (6,6) (6,7) (7,8) (7,9) (8,10) (8,7)

There should be sixteen points on our graph since there are sixteen pairs of numbers, or sixteen legislators. The first member of each pair (in this case the number standing for the number of positive votes) is called the **x-coordinate** and we locate it by using the horizontal scale, or **x-axis.** The second number, or **y-coordinate,** is located by using the vertical scale. When both coordinates have been located, the position of the point has been determined. Here then is the graph.

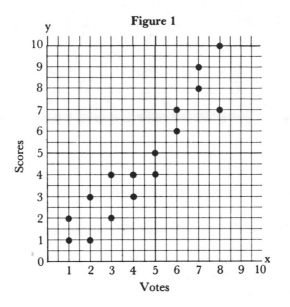

Figure 1

The advantage of graphing the data is that it allows us to see a pattern if one exists. That is, it lets us see how the two sets of measures are related to each other. The particular pattern we are looking for here is a straight line — or rather an approximation of one. As we will see later, this is not the only possible kind of relationship, but it is the only one we can analyze using linear regression, so it is what we are looking for here. And as you can see from the graph, we do have a linear relationship. We can clearly see that as the number of positive votes increases, so do the scores. The relationship is not perfect — there are irregularities — but the trend is apparent. Now in order to analyze this relationship in a useful way, we will have to use some basic algebraic concepts, so in the following section we will consider the relationship between the equation and the graph of a line.

Exercise: Using graph paper, plot the data below with income on the x-axis and candy consumption on the y-axis.

Income: 3, 4, 5, 6, 7, 8, 9, 10, 11, 12, 13, 14, 15
(in thousands)
Candy
Consumption: 6, 7, 9, 8, 11, 10, 13, 16, 17, 14, 20, 26, 26
(in pounds)

Equation and Graph. Look at the following graph of a line.

Example 2:

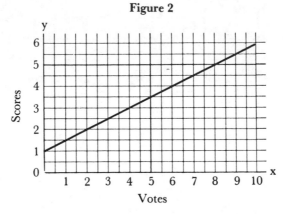

Figure 2

You may remember from algebra that every line has an equation that describes it and no other line, and that every equation can be graphed as one and only one line. The equations for all straight lines have the same general form:

$$y = a + bx$$

When *a* and *b* are replaced by actual numbers, then we have a unique equation that describes a unique line. Here is what the letters stand for:

y = the *y*-coordinate of any place on the line (dependent variable)

x = the *x*-coordinate of any place on the line (independent variable)

a = the **y-intercept** — the place at which the line crosses the vertical scale or *y*-axis. It is the value of *y* when *x* is 0.

b = the **slope** of the line — the rate at which the *y* values change with respect to the *x*'s.

Statisticians call the variable which is being predicted or estimated the **dependent variable,** and the one that is known and is used to make the prediction the **independent variable.** These names mean that what we predict depends on what we know. They should not be taken to mean that the dependent variable is caused by the independent variable. As we shall see later, there may or may not be a cause and effect relationship between the dependent and independent variables.

In order to be consistent, statisticians always use the *y*-axis to plot the dependent variable and the *x*-axis to plot the independent variable. Sometimes we chose one set of data to use as the independent variable because we have more information about it and hope to be able to predict the other set from it. Other times we might really have reason to believe that there is a cause and effect relationship and then we would of course choose as our independent variable the one that we thought was causing the de-

pendent variable. If we are looking at data out of context, as in exercises in
a text book, it is not always obvious why one variable is chosen to be the
dependent and the other independent variable. In Example 1, voting records are more easily obtainable than figures on voters' impressions, and the
former are also more likely to cause the latter, so there is a logic to making
votes an independent variable. In the exercise above, it similarly seems
more likely that we would be investigating the possible effect of income on
candy consumption than the other way round. However, it is conceivable
that a special study might want to investigate the possibility of using
wholesale candy distributors' regional figures as a shorthand way of predicting regional income disparities, in which case it would use the x-axis
for candy and y-axis for income.

Returning now to the equation of a line, we note that if we want to
write an equation for the line in Figure 2, we must find what the values of a
and b are. That is, we must find the values of the y-intercept and the slope
of the line.

The y-intercept can be read directly off the graph. We see that the line
crosses the y-axis (vertical scale) at $y = 1$. So we know that the y-intercept
is $a = 1$. The slope is only slightly more trouble to find. The formula for
finding the slope is:

$$b = \frac{y_2 - y_1}{x_2 - x_1}$$

That is, we take the difference between the y-coordinates of any two points
on the line and divide that by the difference between the x-coordinates of
those same two points. A shorthand way of finding the difference between
coordinates, without actually subtracting one number from another, is to
count the squares on the graph paper. How many squares does the line rise
or fall for every square that it moves from left to right? Since we're interested in **ratios,** not absolute numbers, it doesn't matter how many units
of whatever we're measuring each square represents.

Exercise: Try this method for figuring the slope of the line in Figure 2.
Pick two points on the line; count how many squares the line
rises between them; divide by the number of squares it moves to
the right. Then calculate the slope by picking two other points
and using the formula. You should get the same result.

$$a = 1$$
$$b = \frac{1.5}{3} = \frac{1}{2} = .5$$

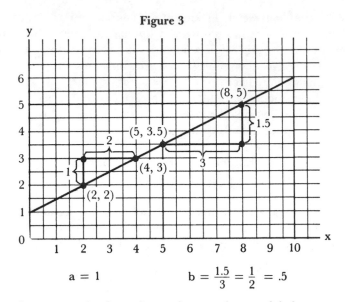

Figure 3

$$a = 1 \qquad\qquad b = \frac{1.5}{3} = \frac{1}{2} = .5$$

If we do not use the formula, we have to be careful that we get the sign of the slope right. The easiest way to do that is to look at the line. If it goes up from left to right, then the slope is positive, or b is greater than 0. If the line goes down from left to right, then the slope is negative, or b is less than 0. The line in Example 2 has a positive slope.

Exercise: Using Figure 3, pick some other numbers for x. Find the corresponding y's on the y-axis. Then see if you get the same slope: $b = .5$.

Having found that the slope of our line is $b = \frac{1}{2}$ or .5, and that the y-intercept is $a = 1$, we can substitute these values into the general equation and get

$$y = 1 + .5x$$

We can check to make sure the equation is correct by picking any number for x and seeing if we get the corresponding y when we substitute it into the equation and solve for y. Let's use $x = 3$. Then we have

$$y = 1 + .5(3) = 2.5$$

If we look at the graph, we see that $(3, 2.5)$ is actually on the line.

A few general remarks about the slope and y-intercept may be useful. The slope tells us about the angle of the line with respect to the vertical or horizontal axis. All the lines in the graph below have the same slope, but each has a different y-intercept.

Figure 4

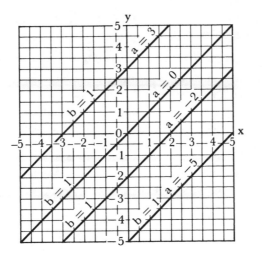

By contrast, each of the lines in the graph below has the same *y*-intercept but a different slope. If the slope of a line is negative, the line will go from the top of the left side of the graph to the bottom of the right side, while if the slope is positive, the line will go from the bottom of the left side to the top of the right. Since the slope tells us how fast *y* (the value on the vertical scale) changes with respect to *x* (the value on the horizontal scale), a higher slope means a steeper line. A horizontal line has a slope of $b = 0$. Vertical lines have no numerical slope because for a vertical line the difference between any two *x*'s is zero and we cannot divide by zero.

Figure 5

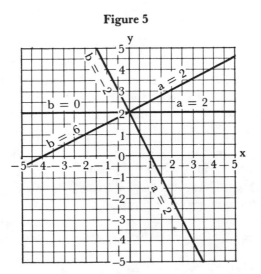

We have seen that we can write the equation of a line when we have the graph by reading off the value of the y-intercept and calculating the slope from any two places on the line. Now let us see how we can graph a line when we are given the equation.

Example 3:   Draw the graph of the equation $y = 4 + 3x$.

We know from elementary geometry that it only takes two points to determine a straight line. We can find those two points by substituting any number for x in the equation and then seeing what the value of the corresponding y is. Let's use $x = 0$ and $x = 2$. Then we get:

$$y = 4 + 3(0) = 4 \quad \text{and} \quad y = 4 + 3(2) = 4 + 6 = 10$$

Therefore the two points are (0, 4) and (2, 10). When these two points are connected by a straight line, the graph looks like this.

Figure 6

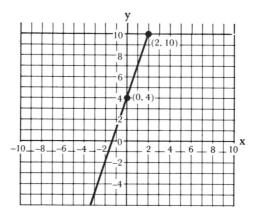

It is usually a good idea to check the accuracy of your graph by calculating a third point to see if it falls on the line. If not, then you have made an error in calculating one of the three sets of coordinates and should check further.

Exercise: Go back to Example 3 and now try $x = 1$ in the equation. What do you get for y? Does the point lie on the line?

   Line of Best Fit. We can now apply our knowledge of graphs and equations of straight lines to the problem of analyzing the relationship between two sets of measures. Let's return to the scatter diagram from Example 1. We can see that the pattern of points vaguely suggests a line which starts in the lower left hand corner of the graph and proceeds to the upper right hand corner. Let's try to draw a line that "fits" the pattern of points as well as possible. We might make several attempts and pick the one that seems to be the best fit.

Figure 7

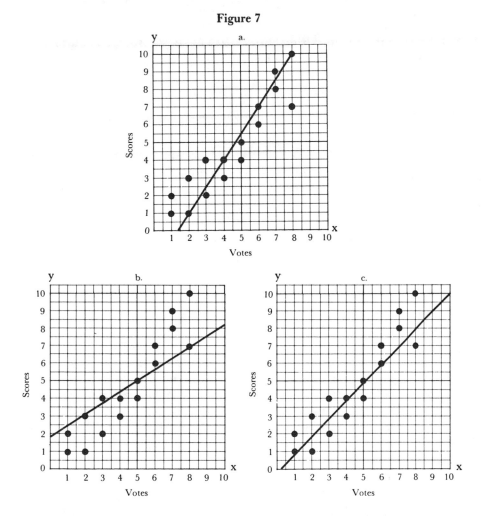

Do you agree that the last try comes closest to our intuitive notion of a line of best fit? Without worrying too much at this stage about how we get the line of best fit, let's see if we can write an equation for the line we just got. In order to write the equation, we need to find a and b. We can see that the line intercepts the y-axis just below the zero point. We might estimate the y-intercept to be $a = -.25$. To calculate b we can pick any two places on the line. Let's use $(.5, 0)$ and $(5, 5)$. Then

$$b = \frac{5 - 0}{5 - .5} = \frac{5}{4.5} = 1.11$$

Substituting these values of a and b into the general equation, we get

$$y = -.25 + 1.11x$$

This is the equation for the line of best fit for the data in Example 1. Or rather, it is an approximation of that line. In statistics, the line of best fit is known as the **regression line.** Before we discuss more accurate methods of determining the regression line, let's see how we can use it.

First of all the regression line has a positive slope (1.11) which tells us that as the numbers in one set of data get larger, so do the numbers in the other. That is, the more positive votes a legislator casts, the higher that legislator's rating is with women voters. Secondly, the equation allows us to make estimates, or predictions from one set of data to the other. Notice that we cannot predict from a scatter diagram but we can from a regression equation. For example, suppose a legislator voted positively on 9 bills. What would we predict her sensitivity score on the voter survey to be? We can answer that question by substituting 9 for the x in the regression equation and calculating the value of y:

$$y = -.25 + 1.11(9) = 9.74$$

We can think of the regression line, then, as an approximation of the original data that allows us to determine what the general relationship is between the two sets of numbers and to predict or estimate one set from the other.

Calculating the Regression Line. Now that we have at least an intuitive understanding of what the regression line is, we must look at a more exact way to calculate it. Although the derivation of the formulas used are beyond the scope of this book, we can get an idea of what is involved by going back to the definition. We said that the regression line was the line that best fit the points. We can make this idea of "best fit" more precise by measuring the distance from each point on the graph to the line. Notice that the distance is measured vertically. In the graphs below, the distances between the points and the line are shown. Do you see that the distances are smallest in the third graph?

Figure 8

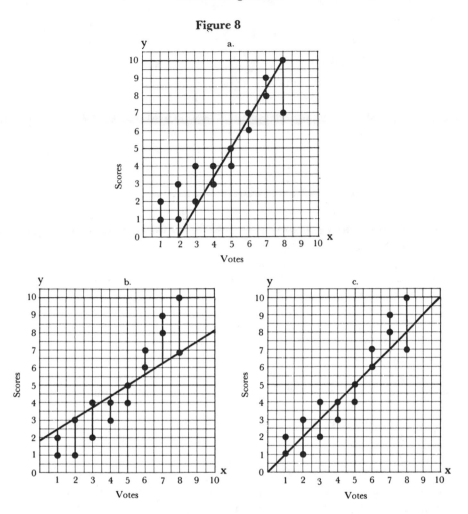

Mathematically, the line of best fit is defined as the line for which the sum of the squares of the distances is smallest. We can write a formula for this sum:

$$D^2 = \Sigma[y - (a + bx)]^2$$

The y in the formula represents the actual y value of the point. The $(a + bx)$ represents the y-coordinate of the place on the line which corresponds to that y (that is, which is directly above or below the point). So $y - (a + bx)$ is the vertical distance between the point and the line. When we square each of these distances and add them up, we have D^2, the sum of the squares of the distances. When we apply further mathematical procedures, we can calculate values for a and for b which reduce D^2 to its smallest possible value. This procedure gives us the following formulas:

$$b = \frac{n(\Sigma xy) - (\Sigma x)(\Sigma y)}{n(\Sigma x^2) - (\Sigma x)^2} \text{ and } a = \frac{\Sigma y}{n} - b\frac{\Sigma x}{n} \text{ or } a = \bar{y} - b\bar{x}$$

When we want to calculate the regression line as accurately as possible, then, these formulas provide the means for doing so.

The second formula, the one for a, also provides us with a hint as to how to make a more accurate "eyeball" estimate from the scatter diagram. Look at the final version of that formula expressed in terms of the mean of the x-coordinates and the mean of the y-coordinates. (In Example 1, that would be the mean of the voting records and the mean of the survey scores, respectively.) We can change the form of that equation by adding $b\bar{x}$ to both sides, giving us: $a + b\bar{x} = \bar{y}$, or $\bar{y} = a + b\bar{x}$. Since $y = a + bx$ is the general equation which locates points on a straight line, we have just learned that the point $(\bar{x}, \bar{y})$ is located on the regression line.

If we stop to think about it, forgetting all the mathematics, this is not a surprising conclusion. It is quite reasonable that the line of best fit would go through the place whose coordinates are the mean of the x's and the mean of the y's. The means are, so to speak, in the center of our numerical data, so it is logical that they are in the center of our scatter diagram as well. Knowing this makes our eyeball estimate more exact, since we can begin by locating this point $(\bar{x}, \bar{y})$, and choose the line passing through it which seems to give the best fit. That is, we can place a ruler on that point and rotate it, looking for the best fit with all the points on our scatter diagram.

Let's look at another example and see how we can get both a better eyeball estimate and an exact calculation of the regression line.

Example 4: Suppose we want to compare years of education with salaries earned. We use a random sample of fifteen individuals and get the following data.

Individual:	A	B	C	D	E	F	G	H	I	J	K	L	M	N	O
Education: (in years)	9	12	12	12	14	14	15	16	16	16	17	17	18	20	20
Salary: (in $1000)	12	6	7	9	15	7	11	8	12	14	9	17	13	10	18

First we calculate the mean of the x's and y's to give us a reference point. We find that

$$\frac{\Sigma x}{n} = 15.2 \quad \text{and} \quad \frac{\Sigma y}{n} = 11.2$$

We locate that point on our graph, and then try to adjust the angle of the line through that point to best fit our data. We may end up with something like the graph below.

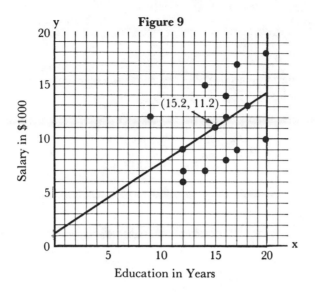

Figure 9

From this graph we can find values for a and b. We see that $a = 2$. To calculate b we pick the points at each end of the line and calculate $\frac{14 - 2}{20 - 0} = 12/20 = .6$. We get the equation $y = 2 + .6x$ for our regression line.

In order to calculate the regression line accurately, however, we must use our formulas for a and b. In that case, our calculations will look like this:

Table 1

DATA

x	y	xy	x^2
9	12	108	81
12	6	72	144
12	7	84	144
12	9	108	144
14	15	210	196
14	7	98	196
15	11	165	225
16	8	128	256
16	12	192	256
16	14	224	256
17	9	153	289
17	17	289	289
18	13	234	324
20	10	200	400
20	18	360	400
168	228	2625	3600

$n = 15$
$\Sigma\, x = 15$
$\Sigma\, y = 168$
$\Sigma\, xy = 2625$
$\Sigma\, x^2 = 3600$
$(\Sigma\, x)^2 = (228)^2 = 5198$

$$b = \frac{n(\Sigma\, xy) - (\Sigma\, x)(\Sigma\, y)}{n(\Sigma\, x^2) - (\Sigma\, x)^2} = \frac{(15)(2625) - (228)(168)}{(15)(3600) - (51984)}$$

$$= \frac{1071}{2016} = .53125$$

$$a = \frac{\Sigma\, y}{n} - m\frac{\Sigma\, x}{n} = 11.2 - (.53125)(15.2) = 3.125$$

From Table 1 we found $a = 3.1$ and $b = .53$. Rounding off these numbers, we get the equation:

$$y = 3 + .5x$$

We can compare the accuracy of the two regression equations by computing D^2 for each.

Table 2

DATA		D^2 when $y = 2 + .6x$			D^2 when $y = 3 + .5x$		
x	y	$2 + .6x$	$y - (2 + .6x)$	$[y - (2 + .6x)]^2$	$3 + .5x$	$y - (3 + .5x)$	$[y - (3 + .5x)]^2$
9	12	7.4	4.6	21.16	7.5	4.5	20.25
12	6	9.2	−3.2	10.24	9.0	−3.0	9.00
12	7	9.2	−2.2	4.84	9.0	−2.0	4.00
12	9	9.2	−0.2	0.04	9.0	0.0	0.00
14	15	10.4	4.6	21.16	10.0	5.0	25.00
14	7	10.4	−3.4	11.56	10.0	−3.0	9.00
15	11	11.0	0.0	0.00	10.5	0.5	0.25
16	8	11.6	−3.6	12.96	11.0	−3.0	9.00
16	12	11.6	0.4	0.16	11.0	1.0	1.00
16	14	11.6	2.4	5.76	11.0	3.0	9.00
17	9	12.2	−3.2	10.24	11.5	−2.5	6.25
17	17	12.2	4.8	23.04	11.5	5.5	30.25
18	13	12.8	0.2	0.04	12.0	1.0	1.00
20	10	14.0	−4.0	16.00	13.0	−3.0	9.00
20	18	14.0	4.0	16.00	13.0	4.0	16.00
				153.2			149.00

So we see that $D^2 = 153.2$ for the estimated equation while for the equation calculated with formulas we get a slightly better $D^2 = 149$. Notice that D^2 is useful for comparing different candidates for the line of best fit for the same data. It is not useful for judging how well the line of best fit describes the particular data. For a measure of that, we must use correlation, which will be discussed in the next chapter.

Using the calculated regression equation, $y = 3.1 + .53x$, we can see that the relationship between the two sets of data is such that as one measure increases the other increases. It is not surprising that the more education people have, the higher their salaries tend to be. We can also use the equation to estimate missing values. For example, none of the people in the sample had exactly 13 years of education. If we substitute 13 for x in the equation, we can calculate the salary to be expected for someone with 13 years of education.

$$y = 3.1 + .53(13) = 9.99$$

So we would expect someone with 13 years of education to earn approximately $9,990. But remember, we have no measure of how accurate that estimate is likely to be. At this stage we must take it as a rough approximation. Only the graph itself, showing the distances between points and the regression line, gives us an intuitive sense of how rough that estimate is likely to be.

PART B

Since regression is so closely related to correlation, we will wait until the next chapter to discuss certain problems involved in applying regression analysis to actual data — particularly questions having to do with accuracy. However, there are other problems having to do with the application of regression analysis which we can discuss now.

Linear and Multiple Regression. In Part A we talked about linear regression, that is, an analysis which takes into account the effect of only one variable on another. In real life situations, things are seldom so simple. Take for example the relationship between education and salaries. We know from experience that there are a great many factors that influence income other than education. Christopher Jencks, in his latest book (*Who Gets Ahead: Determinants of Economic Success,* Basic Books, 1979), has pointed out the importance of the opportunities provided by family background, school ties, and just plain luck, along with a host of other factors. Linear regression therefore is not the most useful tool for analyzing so complex a situation. There exists, however, another statistical tool, called **multiple regression,** which is capable of handling a number of different independent variables at the same time. Although the study of multiple regression is beyond the scope of this book, it is important to realize that such a technique exists, and that it is of great importance in analyzing complex situations.

Prediction vs. Estimation. It is customary to speak of the use of regression lines to predict data. In some sense this is quite misleading. When we use the term "predicting," we usually have in mind anticipating future events. We think of predicting elections, future population growth, or inflation rates. Yet if we are talking about predicting some measure which we analyze against time, predicting the future is not a legitimate use of regression analysis. One of the limitations on regression is that the range of prediction is limited to the range of data. That is, if we have population figures for the years 1968 to 1978 and find that there is an increase or decrease over that period of time, we can use the regression line to "predict" what the population should be for any year *between* 1968 and 1978 but not for anytime before 1968 or after 1978. The reason for this limitation should be obvious when we think about it. Just because the relationship between two measures is linear within certain bounds does not mean it is linear everywhere. Consider the growth of a child. If you plot the age of an infant against its weight for the first year of life, a fairly reasonable linear relationship is obtained. But imagine how misleading it would be to continue that line for the life span of the individual!

This limitation on predicting beyond the range of data applies not only to data over time. Consider the following example.

Example 5:   Recently there has been much talk about saving fuel. The following graph shows one estimate of how much more or less fuel is needed when thermostats are set above or below 70.

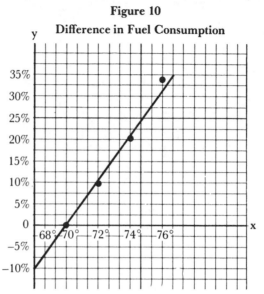

Figure 10

Difference in Fuel Consumption

Although it is safe to use this graph to predict what fuel savings might be at 69° or what increases might be necessary at 73°, predictions far outside of the range of the graph are in danger of being quite wrong. At the upper end of the scale, the furnace might become so inefficient that the increase would be much greater, while at the lower end, the relationship between inside and outside temperatures might be influenced by such factors as insulation and day-night temperature fluctuations, making general predictions uncertain at best.

The Regression Fallacy. The term "regression" was first used by Sir Francis Galton, an early statistician, in a paper published in 1885. This paper has the distinction of both introducing the important concept of regression analysis and committing the first and most fundamental regression fallacy. The phenomenon which Galton observed and described quite correctly is one we still call "regression toward the mean." Galton, in his study of genetics, noted that tall parents tended to have tall offspring — but not as tall as themselves. Similarly, short parents tended to have short offspring, but again not as short as themselves. That is, the heights of the offspring tend to regress toward the mean. This is a well known phenomenon, and one recognized in a variety of situations. For example, to the extent that extremely tall or short people result from a rather unusual combination of hereditary and environmental factors, it is not surprising that more likely than not, this rare combination will not be exactly reproduced in the children and so the children will tend to be less extreme specimens. So far so good. The fallacy, however, comes in assuming that therefore there is a general trend toward "mediocrity" in the population. Galton noticed that when we try to predict certain characteristics in children on the basis of those characteristics in parents, we find them to be less extreme in the children. He took this as evidence that over time those characteristics would tend to move closer and closer to the mean. What he failed to take into account, however, was that while very tall parents might have children shorter than themselves on the average, it was also true that a certain number of medium sized parents would, by chance, have very tall children. The loss of tallness from one source would be balanced by a gain from another — that is the nature of random fluctuation. Over time then, the balance between extreme values and mean values does not change.

Regression toward the mean as a statistical phenomenon appears in many areas where one set of scores is used to predict another. One example involves the development of individual human intelligence.

Example 6: In *Inequality,* Christopher Jencks explains how this works. Although early childhood IQ scores are poor predictors of adult competence, those of 4 or 5 year olds do better.

"That means that 4 year olds with IQs of 120 typically have adult scores around 110. Similarly, 4 year olds with scores of 70 have an average adult score of 85. . . . a relatively advantaged 4 year old will typically retain only half this advantage into adulthood, and a low-scoring child will typically overcome half his disadvantage."

Thus, over a period of time, there is a tendency for individual intelligence to regress toward the mean. Jencks goes on to point out, however, that this does not imply a population of mediocre adults:

"This does not mean that there will be fewer adults than children with very high or very low IQs. While those who start out high or low will usually regress toward the mean, their places will be taken by others who started closer to the mean." (pg. 59)

SUMMARY

PART A

Linear regression analysis is a method of examining the relationship between two sets of measures. It is used to see what the nature of the relationship is and to predict the measures in one set from those in the other. The data is best displayed in **scatter diagrams** which are basically point graphs. A straight line which best approximates the data is then found.

The general equation for any straight line is: $y = a + bx$, where y is the y-coordinate of any point on the line, x is the x-coordinate of that same point, b is the slope of the line and a the point where the line intersects the vertical (y-) axis. The formula for the slope is $b = \dfrac{y_2 - y_1}{x_2 - x_1}$ where (x_1, y_1) and (x_2, y_2) are any two points on the line. Any line has one and only one equation that describes it and any equation describes one and only one line. An equation describes a line when the coordinates of any place on the line make the equation true.

The regression line for any set of data is the **line of best fit.** The line of best fit is the line for which the sum of the squares of the vertical distances from the data points to the line is as small as possible. This condition is met when the following formulas are used to determine a and b:

$$b = \frac{n\Sigma\, xy - \Sigma\, x\, \Sigma\, y}{n\Sigma\, x^2 - (\Sigma\, x)^2}, \qquad a = \frac{\Sigma\, y}{n} - b\frac{\Sigma\, x}{n} \qquad \text{or} \qquad a = \bar{y} - b\bar{x}$$

The regression line is used as an approximation of the data points. It tells us what the relationship is between the two sets of data and allows us to predict or estimate one set from the other. When any number is substituted for x in the regression equation $y = a + bx$, then the corresponding y can be calculated.

PART B

There are two problems which arise when regression is used to predict data. The first is that regression cannot legitimately be used to predict beyond the range of the data. Therefore it may be more accurate to use the term "estimate" instead of "predict."

Secondly there is the problem of regression toward the mean. This means that in a randomly distributed measure, predictions from extreme values can be expected to be less extreme. However, this cannot be interpreted as a general trend toward the mean, since this trend will be balanced by the occasional mean value that projects to a more extreme value.

Problems dealing with the accuracy of prediction and with causal relations between the two sets of data will be discussed in the next chapter.

EXERCISES

PART A

1. *Make graphs for the following equations:
   a. $y = 6 + 4x$
   b. $y = 2 - .5x$
   c. $y = -8 + \frac{1}{3}x$

2. Write equations for the following lines:

a.

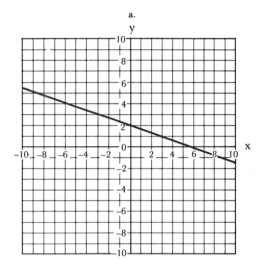

b.

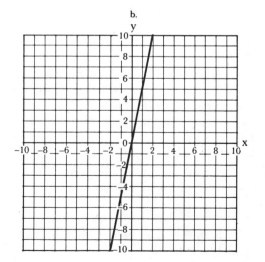

c.

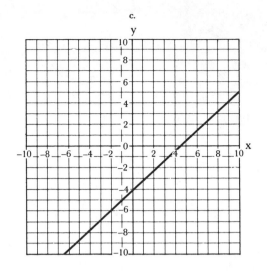

Use the following data for question #3:

Ten students were given a test in manual dexterity and a test in spatial perception, with the following results:

Dexterity score: 3 4 4 5 5 5 6 7 8 9
Spatial perception: 2 5 4 6 3 7 7 8 5 8

3. Plot the data on a graph, putting the dexterity scores on the *x*-axis (horizontal scale) and the spatial perception scores on the *y*-axis (vertical scale). Be sure to make your graph large enough.

a. *Draw a "line of best fit" by what looks reasonable, using the point

$$\left(\frac{\Sigma x}{n}, \frac{\Sigma y}{n} \right).$$

b. Draw vertical lines from each point to the line. Count squares of graph paper to see what the sum of the distances is.

c. Try one or two more lines to see if you can do better. Check by counting the distances from the points to the line.

d. Use your line to estimate the spatial perception score for someone who scores 2 on the dexterity test. For someone who scores 8.5.

e. Use the formulas to find the line of best fit. Recalculate the predictions in *d* if the line is different.

PART B

In which of the following cases can regression analysis properly be used? Explain why and how.

1. From the *Boston Globe* of June 18, 1978:

> Following the postwar 'baby boom' years of the 1950s and 60s, West Germany reached an all-time population high of 62.1 million in the early 1970s. But since 1973 annual deaths have exceeded births by around 100,000 a year, so that the figure is now down to less than 61.4 million, including West Berlin and some 1.9 million foreign laborers, accompanied by around 1.5 million dependents, neither of whom count as immigrants or Germans.
>
> By the year 2028, the demographers estimate, the population will have declined to around 44 million, and by 2050 to less than 37 million. (pg. A3)

2. *From *Human Nature,* May, 1978:

> The accompanying table shows, by way of example, the relation between liver cancer rates and estimated average intake of aflatoxin, the carcinogenic product of a mold that has concerned the FDA because of its prevalence in corn grown in the Southeast in 1977.

Location	Yearly cancer rate per 100,000 people	Daily aflatoxin intake (ng/kg of body weight)
Kenya—high altitude	0.7	3.5
Thailand—Songkhla	2.0	5.0
Kenya—middle altitude	2.9	5.8
Kenya—low altitude	4.2	10.0
Thailand—Ratburi	6.0	45.0
Mozambique—Inhambane	25.4	222.4

3. The recent decline in SAT scores among high school graduates can be explained by the concept of regression to the mean. Over the years, fewer and fewer people have had extremely high scores because of the statistical tendency to regress toward mediocrity and so the average of all scores has declined.

Chapter 14

CORRELATION

In the last chapter we developed the idea of the regression line as a method for predicting one set of data from another. In Example 4 we considered the relationship between numbers of years of schooling and numbers of thousands of dollars earned for 15 people. We plotted these values on a graph, using one point for each person. Then we found the line which best approximates the set of points. We noted that we could use this line to predict the annual income from the number of years of schooling a person had completed. But we left some unanswered questions. Although we found we could make a prediction, we had no measure of how good that prediction was likely to be. When we use a regression line to make a prediction, we do so because we believe that there is a significant relationship between the two sets of data. So far we have not considered any way of testing that belief. Nor have we said anything about the cause of the relationship between the two sets of data. These are both difficult problems.

The first problem can be solved at least to the extent of calculating probabilities. The second raises the question of cause and effect, and our discussion will call attention to the kinds of problems raised by these considerations, rather than presenting solutions.

PART A: The Correlation Coefficient

We will now develop the idea of the **correlation coefficient,** a number which measures the correlation between two sets of data. The word "correlation" means just what it says — co-relation — mutual relationship. So the correlation coefficient measures the extent to which a relationship exists between two sets of data. When we plotted the data about education and earnings, the graph seemed to indicate that a relationship existed. More education was generally related to higher salaries. The correlation coefficient is a number which measures the extent to which this is true. And by doing that, it gives us a measure of the accuracy of any prediction we might make about how much money one is likely to earn after a certain number of years of schooling.

Example 1:

Individual:	A	B	C	D	E	F	G	H	I	J	K	L	M	N	O
Education: (in years)	9	12	12	12	14	14	15	16	16	16	17	17	18	20	20
Salary: (in $1000)	12	6	7	9	15	7	11	8	12	14	9	17	13	10	18

Figure 1

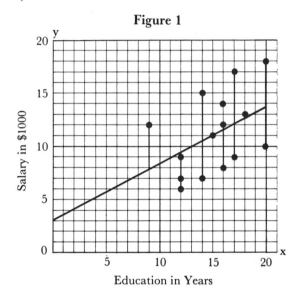

Each point represents the number of years of education and the salary of one individual; the diagonal line is the line of regression, which we found to have the equation:

$$y = 3 + .5x$$

The vertical lines indicate the distances from each point to the regression line. Remember that we defined the regression line as the line of best fit — the line for which the sum of the squares of these vertical distances was as small as possible. We expressed the sum of the squares using the formula:

Formula 1: $$D^2 = \Sigma \, [y - (a + bx)]^2$$

where D^2 stands for the sum of the squares when y is the actual y-coordinate of each point, in this case the actual salary of each individual, and the $(a + bx)$ stands for the y-coordinate of the corresponding place on the line whose equation is $y = a + bx$. (By "corresponding place" we mean the point on the line which has the same x-coordinate as the actual data point and so represents, in this case, the salary predicted by the regression line for that individual.) The difference between the actual and the predicted salaries is represented by the vertical length of the line joining the point to its corresponding place on the regression line. If we then square each of these differences and add them all up, we have the sum of the squares of the differences between the actual and the predicted y values, or D^2.

This number, D^2, is in itself a rough measure of how well the regression line predicts the data. If D^2 is small, it means that the differences between predicted and actual salaries is slight, so the predictions are good. If, on the other hand, D^2 is large, we know that there is a big gap between predicted and actual values, so the predictions are poor. The difficulty lies in defining what we mean by "large" and "small," "good" and "poor." How do we interpret the fact that in the current example $D^2 = 149$?

Part of the difficulty lies in the fact that the size of D^2 is affected by two factors which have no corresponding effect on the value of the prediction. The first of these is the number of cases we have. In our example, we looked at 15 individuals. Suppose we increased the number of individuals to 20. Unless each of the 5 new points representing the 5 new individuals happened to fall exactly on the regression line, D^2 for this group of 20 would have to be larger than D^2 for our original group of 15. But that larger number would not necessarily represent a worse prediction. Secondly, the size of D^2 is affected by the size or range of numbers in our data. The salaries we are measuring fall between \$6,000 and \$18,000. Suppose instead of our group of 15 beginning workers, we had looked at a group of executives with salaries between \$30,000 and \$90,000.

Example 2:

Individual:	A	B	C	D	E	F	G	H	I	J	K	L	M	N	O
Education:	9	12	12	12	14	14	15	16	16	16	17	17	18	20	20
Salary:	60	30	35	45	35	75	55	40	60	70	45	85	65	50	90

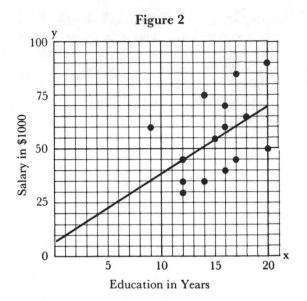

Figure 2

Although the graphs in Examples 1 and 2 look alike and seem to show the same degree of relatedness between the two sets of data, if we calculate D^2 for the executives, we get $D^2 = 3812$, a number much larger than D^2 for our original group.

We can see that this is the same kind of difficulty we encountered with raw scores or with ranks (Chapter 2). There we solved the problem by forming a ratio, in the first case between the raw score and the number of items, in the second case between the rank and the number of scores. In our present predicament it is a little more difficult to see how to form our ratio. If we divide D^2 by the number of points on our graph, we don't deal with the problem of scale; if we divide by the highest value of y, or the mean value, or any other measure meant to deal with scale, we don't deal with the number of cases. The solution is rather ingenious, if a little complicated.

The solution is to compare D^2 to a special measure we will call A^2, which is the sum of the squares of the distances from our points on the graph to a line which would represent *no* correlation between our two sets of data. This solves our numerical comparison problem, since A^2, like D^2, is affected both by the number of points on the graph and by the scale. But more importantly, it solves the logical problem of how to discuss the significance of a linear relationship. By comparing D^2 to A^2, we are asking, "How much better is our regression line as a predictor than a line which is no use at all?"

Now, what do we mean by a line which represents no correlation? Remember that in the last chapter we said it was always possible to find a line of best fit, no matter how little relationship actually existed between the two sets of data. The formulas will give us values for a and b no matter what, and from those values we can graph the line $y = a + bx$. Suppose, for instance, that we asked a sample of individuals to tell us the time shown on their watches and the number of children they had. If we used the times for our x's and the numbers of children for our y's, we would get a regression line which had a slope (b) very close to 0 and which crossed the y-axis (point a) at approximately the mean of all the y values. If the sample were infinitely large, we would get exactly $b = 0$ and $a = \bar{y}$.

In such an example, changes in x (time) would have no effect on y (children). The best prediction for the y-score corresponding to any x-score is $\bar{y}$ (the mean number of children reported). So the equation of the regression line, when there is no relationship between the two sets of data, is:

$$y = \bar{y} + 0x, \text{ or simply } y = \bar{y}$$

Now we can return to the problem of calculating a number, A^2, which we can use with D^2 to form a ratio which will measure correlation. Let's begin by taking our graph of salaries and education (Example 1) and drawing on it the line $y = \bar{y}$. If we add up all the salaries our 15 subjects make and divide by 15, we find that the mean of the y's is $\bar{y} = \$11,200$. We can then draw the line $y = \$11,200$ on our graph and mark the distance from y to $\bar{y}$ for each point. Compare the two graphs below, one with the distance from y to $\bar{y}$ indicated for each point, and the other with the distance from y to $(a + bx)$, in this case $(3.1 + .53x)$, indicated for each point.

Figure 3

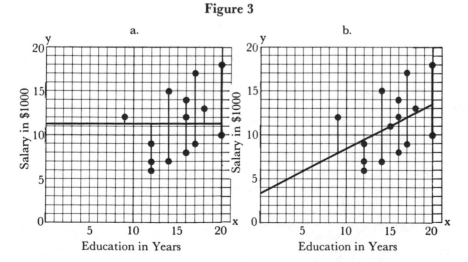

As we might expect, the distances to the line representing no relationship are greater than the distances to the line of best fit. The formula for the squares of these distances is:

Formula 2: $$A^2 = \Sigma (y - \bar{y})^2$$

where y is the y-coordinate of each actual point on the graph
while $\bar{y}$ is the corresponding point on the line $y = \bar{y}$.

D^2 and A^2 will be equally affected by the size of the y values as well as by
the number of points on the graph. Therefore by forming the ratio D^2/A^2
we solve both numerical difficulties simultaneously.

Now let's see what this ratio will actually look like for our present
example. In Chapter 13 we calculated that $D^2 = 149$. If we use the same
method to calculate A^2 we find that $A^2 = 190$. Then:

$$\frac{D^2}{A^2} = \frac{149}{190} = .78$$

Now let's think for a minute about what values D^2/A^2 might be expected to have in general. Since D^2 is derived from the line of best fit (the
line to which the data points are closest), it will always be smaller than or
equal to A^2. When the correlation between two sets of data is perfect, there
will be no distance between any of the points and the line of best fit; D^2
will be equal to 0, and so will D^2/A^2. When there is no correlation between
two sets of data, D^2 will be the same as A^2, and D^2/A^2 will be equal to 1. In
all other cases the value of the ratio will be some number between 0 and 1.

The only problem with this measure is that it runs counter to expecta-
tions, since generally we would assume that 0 meant no correlation and 1
meant a perfect correlation. We can make a simple adjustment which will
accomplish this change. If we subtract the fraction D^2/A^2 from 1, then the
values will run the opposite way. Suppose the fit of the points and line is
perfect, so $D^2/A^2 = 0$. If we then subtract D^2/A^2 from 1, we have $1 - 0 =$
1. Alternatively, if there is no correlation so that $y = \bar{y}$ is truly the line of
best fit, then $D^2 = A^2$ and $D^2/A^2 = 1$. Subtracting from 1, we get $1 - 1 =$
0. We now have our values running the way we want, with a perfect correlation equal to 1 and no correlation equal to 0. Of course it must be understood that the actual values of 0 and 1 are ideals which are seldom if ever
realized in actual data.

The final step in the calculation of the correlation coefficient is to take
the square root. Although we cannot go into the necessity for this step
here, it should be intuitively plausible that since we originally squared the
distances we will now take the square root. The formula is:

Formula 3: $$r = \sqrt{1 - \frac{D^2}{A^2}}$$

When we take the square root of a number we get two answers, a
positive and a negative one. The square root of 4, for instance, is both 2

and -2; either one, when multiplied by itself, equals 4. When we did standard deviations, we had no need for a negative answer — we could not interpret it — so we simply disregarded the negative answer and used only the positive. When calculating correlations, however, we find that we will sometimes use the negative one, and sometimes the positive one. A **positive correlation** represents the situation where the regression line has a positive slope and therefore the numbers in one distribution increase as the numbers in the other do. A **negative correlation** is one where the regression line has a negative slope, indicating that the values of one distribution increase as the other's decrease. A correlation of $+1$, like a correlation of -1, means that there is perfect correspondence between the two distributions and one can be predicted from the other with 100% certainty. But in the case of the $+1$ correlation, the numbers in each set of data increase together, while in the case of a -1 correlation, one set of numbers will increase while the other decreases.

Exercise: One study finds that the correlation coefficient between cigarette smoking and lung cancer is $r = +.72$. Explain the meaning of r.

A survey found that the correlation coefficient between the price per pound of hamburger and demand for hamburger is $r = -.62$. Explain the meaning of r.

Now let us complete the calculation of the correlation coefficient for our example. We found that the value of $D^2/A^2 = .78$. When we substitute this into the formula for the correlation coefficient, we get:

$$r = \sqrt{1 - .78} = \sqrt{.22} = \pm .47$$

We always round off correlations to two decimal places. Then we must choose the correct sign. Since our regression line had a positive slope — that is, people's salaries tended to increase as the years of education increased — we choose the positive sign. The correlation coefficient for salary and education in the case of our fifteen subjects, then, is $+.47$.

In actual practice, when we calculate a correlation coefficient, we do not use the method just outlined. Just as we found that there was a more efficient formula for calculating the standard deviation than the one that led us to understand the concept, there are more efficient formulas for calculating the correlation coefficient without going through the steps of separately calculating A^2 and D^2. Two versions are:

Formula 4: $\quad r = \dfrac{\Sigma\,(xy) - \dfrac{\Sigma\,x\,\Sigma\,y}{n}}{(s_x)\,(s_y)\,(n-1)} \quad$ where s_x is the standard deviation of the x's and s_y of the y's.

Or when a and b are known, we can use the following formula:

Formula 5:
$$r = \sqrt{\frac{a \Sigma y + b \Sigma xy - n \bar{y}^2}{\Sigma y^2 - n \bar{y}^2}}$$

Either of these formulas is quite tedious and should be computed with the use of a calculator.

So far, we have said very little about what the number r actually means. We have said that it is a measure of the relationship between two sets of data and that 0 represents no relationship while 1 represents a perfect correspondence. But what of the values in between — how are we to interpret them? If we are interested in evaluating the relationship between two measures, we must be able to distinguish between a real relationship and one produced by chance fluctuations. Then, once we feel confident we have a real relationship, we will want to know more about it. How much of the variation in one set is caused by the other? We will examine these questions in the next two sections.

Significance of Correlation

We look now at what we can learn from a particular correlation coefficient r that we have calculated from two sets of measures. Will the same value for r always have the same meaning? How can we tell whether the relationship is real or a matter of chance? These questions both relate to the problem of sample size. We have seen in previous chapters that if we take a small sample, then chance fluctuations are more likely to affect our results than if we take a large sample. Not surprisingly, the same is true when we work with correlations. So the answer to the first question is immediately obvious. No, r does not always have the same meaning. A correlation coefficient of .5 might be very meaningful in a large sample and tell us almost nothing about a very small one. The implication for our other question, then, is that the cut-off point between a significant correlation and one that is likely to be the result of chance depends in part on the size of the sample.

Example 3: Suppose I want to check out the popular notion that musical and mathematical ability go together. I round up six friends and give each a test in musical ability and mathematical aptitude. The correlation coefficient for their scores turns out to be .7. To what extent have I proved my point?

Leaving aside the problem of sample selection (are 6 friends likely to be a random sample?) how do we interpret the results? Intuitively we may feel that the sample is too small to be trusted. By comparison, we might feel that a study of 5000 subjects would convince us even if the actual correlation coefficient was much smaller, say .4 or .5.

We now want to take this intuitive feeling and see how we can make mathematical sense of it. We do that by stating a hypothesis we can test. In previous chapters, we have learned to test a number of different hypotheses by using basically the same procedure. We begin by stating a null hypothesis, then set a significance level, calculate a test statistic, find the critical value which corresponds to the significance level on some appropriate table of values, compare the test statistic to the critical value and on that basis accept or reject the hypothesis. Where before we tested hypotheses such as "There is no difference between the mean of the sample and of the population" or "There is no difference between these two distributions," we will now be testing the hypothesis, "This correlation coefficient is not different from 0." This approach changes the question from "How much of a relationship exists between the two sets of data?" to "Does a real rather than a chance relationship exist?" Then we answer the question, not with a "yes" or a "no," but with a probability.

The procedure is exactly the same as the one we used when we calculated the chi-square statistic for two distributions. First we decided what probability of error we could live with. Then we found a chi-square value on the table by looking in the column headed by that probability and along the row with the appropriate number of cells. Finally, we compared the number we found on the table with the number we had calculated from our data. The number from the table was called the **critical value.** If our calculated number was larger than this critical value, we knew we had a significant difference, since the larger the chi-square, the larger the difference. If it was smaller, we knew the difference was not significant.

On page 308 is a table of critical values for r. Let's see what it tells us about our 6 subjects whose musical and mathematical abilities show a correlation of $r = .7$.

Example 4:   Does a correlation of .7 between musical and mathematical ability based on six data points indicate a real relationship?

To answer this question we go through the steps of hypothesis testing one at a time.

1. **Hypotheses:** H_o: There is no difference between the correlation found between musical and mathematical ability and a correlation of 0.

    H_a: There is a difference between the correlation found and a correlation of 0.

2. **Significance Level:** Set $P = .05$ since there is no obvious reason to use a different level.

3. **One-tailed Test:**   We use a one-tailed test when we are only interested in only a positive or only a negative correlation. A two-tailed test is appropriate if we have no reason to expect one or the other and are look-

ing for any kind of relationship. Here we expect a positive correlation, so we use a one-tailed test.

4. **Critical Value:** We find the critical value by looking at the table. We use the column labeled .05 under one-tailed test. Then we find the appropriate degrees of freedom. For χ^2 it was the number of cells minus 1. For the correlation coefficient we use the number of pairs of measures minus 2. So in this case $df = 6 - 2 = 4$. The critical value from the table then is .7293.

5. **Test Statistic:** Our test statistic is the correlation coefficient which we calculated to be .7.

6. **Conclusion:** Since r is a measure of correlation, the calculated r or test statistic must be larger than the critical value for us to have a real relationship. Since our test statistic is smaller than the critical value, we accept the null hypothesis that the correlation is not different from zero, and therefore not a significant indicator of a real relationship.

Notice that our intuitive feeling that it would be very hard to find a correlation with such a small sample proved to be correct. When we are looking at only 6 people it is quite possible to get a high measure of correlation quite by chance. If we had a much larger sample, then even a value of r much smaller than .7 would be taken as indicative of a real relationship between musical and mathematical ability.

Now let's return to our fifteen subjects whose years of education and earned salary we have been analyzing.

Example 5: Is a correlation coefficient of .47 significant for $n = 15$?

1. **Hypotheses: H_o:** There is no difference between the calculated correlation and zero.

   H_a: There is a difference between the calculated correlation and zero.

2. **Significance Level:** .05

3. **One-Tailed Test:**

4. **Critical Value:** .4409

5. **Test Statistic:** .47

6. **Conclusion:** Reject the null hypothesis.

We see in this case, then, that we do have a real relationship between education and salary.

Exercise: Is a correlation coefficient of .5 between earnings per share and price per share of common stock significant for $r = 10$? (One-tailed test, significance level $= .05$.)

Once we know that a relationship is significant, we can ask a new question: just what does it mean to say that there is a correlation of .47 or .86 or any other specific value? To answer this question, we must look at the coefficient of determination.

The Coefficient of Determination

If we determine from the correlation coefficient that a real correlation exists, we then want to know *how much*. Granted that knowing the value of x (in Examples 1 and 5, years of schooling) is of some use in predicting y (salary), we want to know how much use. Do educational differences account for all variation in salaries, half of variation, or what?

For this purpose we turn to the measure r^2, the **coefficient of determination.** This number, when expressed as a percent, tells us what portion of the total amount of variation in one set of data can be accounted for by its linear relationship to the other. In the case of Example 1, for instance, we found $r = .47$, so $r^2 = .47^2 = .22$. Changed to a percent, $.22 = 22\%$. Therefore we can say that 22% of the variation in salary among our fifteen subjects can be accounted for by the differences in the number of years they spent in school.

In order to see why this is so, we must go back to our original formula for r. Since $r = \sqrt{1 - \dfrac{D^2}{A^2}}$, $r^2 = 1 - \dfrac{D^2}{A^2}$. Remember that A^2 stands for the squared distances from all the y's to the mean of the y's and so is a kind of measure of variation of the y's from the mean. The regression line attempts to explain these differences or this variation in terms of a linear relationship between the x's and the y's. Since D^2 stands for the squared distances from all the y's to the regression line, it represents that part of the variation that is *not* explained by the linear relationship. Note that if all the y's fell exactly on the regression line, D^2 would be 0 and so *all* the variation from the mean would be accounted for by the linear relationship. We can see then that the expression $\dfrac{D^2}{A^2}$ is the ratio between the part of the variation in the distribution of the y's that is not explained by the linear relationship and the total variation. In Example 1 that ratio was .78, so 78% of the variation in salaries is *not* related to educational differences. Subtracting 78% from 1 (or 100%) we find that the remaining portion of the total variation (22%) *is* related to the differences in years of schooling. So the coefficient of determination, $r^2 = 1 - \dfrac{D^2}{A^2}$, is a measure of the portion of the total variation of one measure that is accounted for by its linear relationship to another measure.

Example 6: A car salesman finds that there is a correlation of .39 between the number of minutes of TV advertising time he buys and the number of sales he makes in a given month. What percent of the difference in sales is due to TV advertising?

Since $r = .39$, $r^2 = .39^2 = .1521$. So approximately 15% of the increase in sales can be accounted for by advertising on television.

Exercise: According to one study, the coefficient of determination for the relationship between annual food expenditure and annual national income is $r^2 = .78$. Explain what this means.

Now that we know how to calculate the correlation coefficient and coefficient of determination and to interpret the results, we are ready to go on to a discussion of some of the problems which arise when we put this knowledge to use in some real situations.

PART B

In Part A we learned about three mathematical tools for finding numerical expressions for the relationship between two sets of data: the correlation coefficient, the coefficient of determination, and the siginificance test for correlations. Now we must see how these can be used to reach conclusions about questions arising from actual problems encountered in real life. One of the most difficult research problems confronting everyone from economist to government official to amateur psychologist is the problem of determining cause and effect relationships. Let's see what light our statistical methods can shed on this problem.

Cause and Effect

As we have pointed out earlier, when we find the correlation between two sets of data, we are not saying anything about one thing being caused by the other. We are simply measuring the degree of relationship between them. Now that we have a more precise way of measuring that relationship, we can look again at what we said about causal fallacies in Chapter 1. Suppose we have two sets of data, one measuring phenomenon A, the second measuring phenomenon B. We calculate the correlation coefficient and find that it is high, so we are curious about the causal relationship between the two phenomena. There are four possible ways to explain the relationship:

1. Phenomenon A causes phenomenon B.
2. Phenomenon B causes phenomenon A.

3. Both phenomena are caused by a third phenomenon, C.
4. There is no real causal relationship, the high correlation being a matter of chance.

Let's look at an example in which we consider correlation and causal relationship.

Example 6:   An article appeared in the *Boston Sunday Globe* on August 29, 1976 (p. A2), linking low achievement test scores and high grades. In part, the article says: "Thus, the knowledge that students can demonstrate on objective tests is declining but the subjective ratings of students by teachers are producing higher grades than at any time on record. And although appearing anomalous at first glance, the two phenomena are inexorably linked." How they are linked is never made entirely clear in the article, though there is some hint that high grades are a cover-up, that "grade inflation at least partially blinded many to the reality of the achievement decline."

Suppose now that we took seriously the possibility of a causal connection between low achievement and high grades. We must, of course, have some actual data to work on. Since none is provided in the article, let's make up our own. For our hypothetical data, we could take a random sample of 50 high schools. For each high school, we could calculate the average score on college entrance exams and the average grade point average for each graduating class over the last 15 years (the time during which this change is reported to have come about.) That would give us 750 data points, a goodly number. Suppose now that we calculate the correlation coefficient for these 750 points and do find a negative correlation. This would be consistent with the *Globe* story since a negative correlation is one where one variable (in this case the grade point averages) increases, while the other variable (here the achievement test scores) decreases. Let's say the correlation we find is − .6.

Now how do we decide which of the four possible causal relationships provides the best possible explanation of our data? One approach is to see which ones can be eliminated. Let's begin with #4.

4. There is no causal relationship. The correlation is caused by chance.

Even with the limited experience we have had with testing the significance of correlations, we probably don't need to consult the table to see that with a sample of 750 data points, .6 must be a significant correlation. That rules out the possibility that the connection is the chance result of too small a sample. Of course there are other explanations for chance correlations that have nothing to do with sample size. For example, the West Coast of the United States has for some time been disappearing into the Pacific Ocean at the rate of approximately 9″ a year. At the same time, the population of California has been increasing at the rate of about .5 million people a year. So if we figured the correlation coefficient for land loss and population increase over the past 20 years, we would get a large negative correlation. But this would be quite meaningless, since obviously the correlation would be the result of two trends which are consistently parallel but have no possible causal connection; land erosion and population increase are quite separate events. It seems safe to assume, however, that the correlation we find between classroom grades and achievement scores is not this kind of gratuitous coincidence. Since there exists the very real connection between the two events — both grades and test scores are methods used for the same purpose, to assess student performance — we may assume there must be some real meaning to the correlation. So let's eliminate the explanation that the correlation is purely a matter of chance and move on.

1. High grades cause low achievement test scores vs.
2. Low achievement test scores cause high grades.

The article seemed to suggest that the high grades were at least partially caused by the low achievement scores in an attempt to cover up the schools' failures. How can we check out this possibility? One way is to analyze our data in terms of time. Which began to happen first — the high grades or the low achievement scores? If we find that one phenomenon precedes the other in time, we can at least eliminate it as an effect. That is, if it turns out that grades began to go up and *then* achievement test scores declined, it hardly makes sense to state that grade inflation was caused by achievement score decline. So we may be able to eliminate one possibility, but that does not prove the other is true. Just because one event precedes another is no reason to conclude that it caused it. Every spring we might notice that the crocuses come up before the tulips do, but that does not lead us to conclude that the crocuses cause the tulips to bloom.

In order to prove that one phenomenon is causing the other, we have to be able to control the variables. That means we have to be able to collect data in a way that keeps track of the separate categories we need. In this case, if we believed that declining achievement test scores caused inflated grades, we would have to collect separate data on groups where achievement did not decline and those where it did. If grades became inflated only in the second group and not in the first, we would have strengthened our case.

Finally, if we had some specific hypothesis about the mechanism of the cause and effect relationship, we could do some actual experiments instead of passively collecting data. Suppose we had found that inflated grades preceded reduced achievement test scores. We might formulate the hypothesis that students made less effort when high grades were easy to achieve. We could test this by matching two groups of students with equal ability and achievement and give one group high grades and the other low. Then, after a reasonable period of time, we could give them achievement tests. If the group given high grades actually scored lower, we would have strengthened our case for that hypothesis. (In reality, of course, such a result is highly unlikely. Educational experience has shown that quite the opposite tends to be true: students whose achievements are appreciated respond by learning more rather than less.) Now let's look at the last possibility.

3. Both high grades and low scores are caused by some third phenomenon.

This is probably the most difficult case to test and can never be completely eliminated as a possibility. The only way to test it is to think of possible events that could cause both phenomena and test them out. One hypothesis might be that the climate in education has changed so that the kinds of learning which are now valued by teachers and students are not the kinds which show up best on achievement tests. Students might be performing very well in these areas and therefore getting high grades from teachers, while performing less well in traditional areas tested on achievement tests. This hypothesis could be tested by studying attitudes of students and teachers toward various kinds of achievements and testing students in areas neglected by standard achievement tests.

It is important to remember that none of these four possibilities are really mutually exclusive. Even possibilities #1 and #2 could both be functioning in a "vicious circle" kind of relationship, where higher grades cause poor achievement test scores which cause even higher grades. Furthermore, it is important to remember that the coefficient of determination tells us just what portion of an effect is caused by the linear relationship. So even if we find strong evidence that high grades cause low achievement (or vice versa) the high grades account for only r^2 or 36% $(-.6 \times -.6 = .36 = 36\%)$ of the variation in achievement test scores. The rest is still to be accounted for in other ways.

High correlations, then, are to be seen not as giving us any direct information about cause and effect relationships, but rather as indications that further investigation might lead to some interesting results.

Deceptive Correlations

Let's look now at low, or zero correlations. Can we always assume that no relationship exists if we end up with a very low correlation coefficient? Not at all. There are, in fact, several ways in which very strong relationships can fail to show up when only the correlation coefficient is calculated.

Non-linear relationships. Remember that the correlation coefficient is based on the regression line and therefore only measures a relationship in which the data falls on or near a straight line. Many relationships, however, form a curve when plotted on a graph, rather than a straight line. Regression analysis cannot predict such data. Other means of analysis are available, but these are beyond the scope of this book. It is enough at this stage to be able to distinguish between cases where there is a linear relationship (straight line), those in which the relationship is non-linear, and those with no relationship.

Example 7: A car is tested for gas consumption at speeds of 30, 35, 40, 45, 50, 55, 60, 65 and 70 miles per hour. The data looks like this:

Miles per hour: 30   35   40   45   50   55   60   65 70
Miles per gallon: 15   20   25   26   27   26   25   20 15

Figure 4

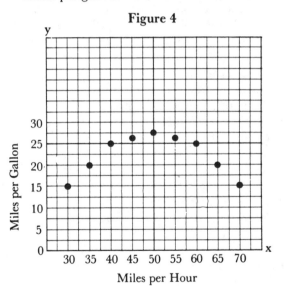

Miles per Gallon / Miles per Hour

Using the calculation formula for the correlation coefficient:

$$r = \frac{\sum xy - \frac{\sum x \sum y}{n}}{(s_r)(s_y)(n-1)} = \frac{9950 - \frac{(450)(199)}{9}}{(13.7)(4.75)(8)} = \frac{9950 - 9950}{520.6} = 0$$

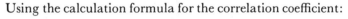

This zero correlation might lead us to believe that there is no relationship between speed and gas consumption. One look at the graph convinces us that quite the contrary is true. It is simply that the relationship is described by a curve, not by a straight line. There are many such relationships that occur in nature and in society. All we can say here is that it is very important to plot data and look at it before deciding that a low correlation means no relationship exists. The graph may show a strong non-linear relationship, as it did in the case of gas consumption.

Inappropriate grouping. Another misleading low correlation comes about when different kinds of data are erroneously grouped together. Here is one such instance.

Example 8: Suppose you are looking at the transportation costs for people living in different parts of Philadelphia. You assume that there will be some correlation between distance traveled and cost of transportation. To your surprise, the data looks like this:

Miles traveled daily: 1   2   2   5   5   7   8   10 11 11
Dollars spent daily: .5 .5 2.2 2.5 .9 .5 2.8 .9 .9 3.1

Figure 5

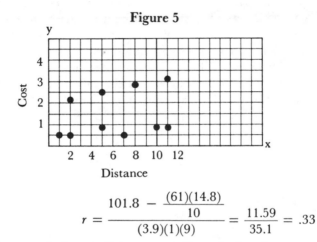

Distance

$$r = \frac{101.8 - \dfrac{(61)(14.8)}{10}}{(3.9)(1)(9)} = \frac{11.59}{35.1} = .33$$

The low correlation coefficient comes from the fact that you are really looking at two different groups of people — one of whom takes the subway to work, paying either $.50 or $.90, while the other group drives to work, paying $2 for parking plus $.10 per mile. Figuring the correlation coefficient for the two groups separately, you would find that the correlation between distance and cost is about .64 for the subway riders, and a perfect 1 for the drivers.

Inappropriate grouping can also cause deceptively high correlations. Consider this example:

Example 9: An industry is being investigated for hazards to employees due to dust inhalation. The industry claims that its workers may have initial problems but quickly adjust to the working conditions, so that recent employees show more absences for respiratory illness than older employees. Their data for a period of ten years looks like this:

Years employed:	1	2	3	4	5	6	7	8	9	10
Average absences:	15	9	8	8	7	6	6	5	4	5

This data yields a correlation coefficient of approximately − .86. This seems to show that the longer people work at this job, the less it affects their health. Does that seem reasonable?

The explanation for this apparent paradox lies in the grouping. The assumption is that we are comparing the same people over a period of ten years, but of course that isn't true. Each year the people who are the most severely affected either quit in disgust, get too sick to work, or die. So the people who have been there for 10 years are members of a select group — the survivors. Lumping them together with those recently hired, and everyone in between, gives a very wrong impression. To really find out what happens over a period of time, it would be necessary to keep track of one group of people for ten years. Monitoring everyone in the group, whether they continued at that work or not, would give a clearer picture of how working conditions affected their health. It is very important, then, in looking at correlations, to be aware of the nature and size of the sample and to see what the actual distribution of data looks like. Then, much caution should be exercised in drawing any conclusions about causal relations between the sets of data being examined.

SUMMARY

PART A

1. The **correlation coefficient,** r, measures the extent to which two sets of data are related. The formula

$$r = \sqrt{1 - \frac{D^2}{A^2}} \quad \text{where} \quad D^2 = \Sigma \, [y - (a - bx)]^2 \quad \text{and}$$

$$A^2 = \Sigma \, (y - \bar{y})^2$$

helps us to understand that we are basically measuring the ratio between the distance of data points from the regression line compared to the distance of data points from the mean of the y's. To calculate r, we can use the following formulas:

$$r = \frac{\Sigma \, (xy) - \dfrac{\Sigma \, x \, \Sigma \, y}{n}}{(s_x)(s_y)(n - 1)} \quad \text{or, where } a \text{ and } b \text{ are known:}$$

$$r = \sqrt{\frac{a \, \Sigma \, y + b \, \Sigma \, xy - n \, \bar{y}^2}{\Sigma \, y^2 - n \, \bar{y}^2}}$$

The correlation coefficient is always a number between −1 and 1.

A correlation coefficient of 1 indicates that there is a perfect linear relationship between the two sets of data — one can be predicted from the other with 100% certainty — and that as one variable increases, so does the other.

A **correlation coefficient of −1** also indicates a perfect linear relationship, but one where as one variable increases, the other decreases.

A **correlation of 0** indicates that no linear relationship exists between the two sets of data.

2. The **coefficient of determination, r^2,** tells us what percent of the total variance of one of the sets of data is accounted for by its linear relationship to the other. We calculate the coefficient of determination by multiplying r by itself and expressing the result as a percent.

3. The **significance** of the correlation coefficient can be determined by using the table of critical values. By comparing our calculated correlation coefficient with the critical value (from the table) which corresponds to the number of data points and to a selected significance level, we can see if the correlation is significant at that level.

PART B

1. **Cause and Effect.** A high correlation coefficient can be explained by one or more of the following possibilities.
   a. Phenomenon A caused phenomenon B.
   b. Phenomenon B caused phenomenon A.
   c. Both phenomena were caused by a third phenomenon, C.
   d. No causal relationship exists, the high correlation being the result of chance.

   Some of these possibilities can be eliminated by careful analysis of the data, by proper grouping of data or by experimentally testing hypotheses.

2. **Deceptive Correlations**
   a. **Non-linear** relationships yield low or zero correlation coefficients even when a strong relationship exists, because the data falls along a curve rather than a straight line.
   b. **Inappropriate grouping** can lead to a **deceptively low** correlation coefficient when two or more groups are combined into one even though each group has distinctive features with respect to the relevant variable.
   c. **Inappropriate grouping** can lead to **deceptively high** correlations when different groups are combined and/or when parts of groups are missing from the total data.

EXERCISES

PART A

Use the following data for all problems in Part A. Assume that ten people are tested on their verbal ability and on manual dexterity. The tests are scored on a scale of 1–10.

Individual:	A	B	C	D	E	F	G	H	I	J
Manual score:	1	2	3	3	6	6	8	8	9	9
Verbal score:	6	8	3	4	10	9	10	2	3	4

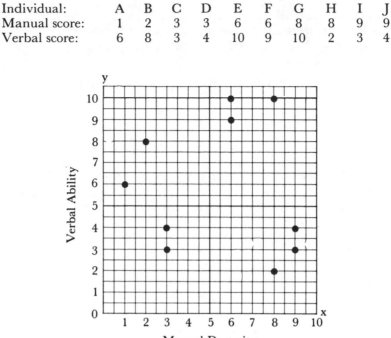

1. *Just by looking at the graph, what would you guess the correlation coefficient to be?

2. If the regression line is $y = 6.45 - 10x$ and $\bar{y} = 5.9$, find the correlation coefficient using the formula

$$r = \sqrt{1 - \frac{D^2}{A^2}}$$

3. *Suppose now that you are interested in seeing if there is any difference between people with high verbal scores and people with low verbal scores in how well their verbal and manual skills correlate. To check this out, calculate the correlation coefficient for the people who scored above 5 on verbal ability. Then do a separate calculation for the people who scored below 5 on verbal ability.

4. Use the table of critical values to see if either (or both) of the correlation coefficients found in #3 above is significant at a .05 level. Explain exactly what your answer means.

5. *Find the coefficient of determination for the correlation coefficients found in #3. Explain what they mean.

6. What conclusions can you draw from your answers to #1–#5 above?

PART B

1. On August 31, 1976 the *Boston Globe* reported on a new study dealing with the connection between heroin addiction and crime. The report comes up with the following interesting conclusion:

> It may be a general need for income, rather than the need for money to buy drugs, that lies behind the criminal behavior of those users who commit crimes against persons and property. In fact the causal relationship may be the other way around; criminality may in fact lead to drug use, as the successful criminal begins to use drugs as a luxury or status symbol.

   Suppose you wanted to test this hypothesis. What kind of data would you collect, what calculations would you perform, and how would you interpret your results?

2. What fallacy can you find in each of the following examples?

   a. An article in the *National Enquirer* of July 27, 1976 is headlined: FINGERPRINTS CAN GIVE EARLY WARNING OF BREAST CANCER and says in part:

   > "We found a bizarre, unusual print called a 'composite' in 24 percent of our breast cancer patients . . . Such a print is only found in 6 people in every 1000 in the general population, but we found it in 240 per 1000 [of people having breast cancer] . . ." From this information the article concludes: "If our results hold up on an international basis, it means we could fingerprint every 5-year-old girl and know right away if she's susceptible to breast cancer later."

   b. *A doctor reports that the older his patients are, the more likely they are to be women. His audience concludes that among older people, women are more sickly than men.

   c. *A psychologist wants to see how students' anxiety affects their performance on tests. He rates their level of anxiety on a scale from 1 to 10 and scores their tests on a similar scale. The data looks like this:

Anxiety index:	2	3	4	5	6	7	8	9
Test score:	6	7	8	9	8	7	7	6

   The correlation coefficient turns out to be $-.11$.

   The psychologist concludes that there is no connection between anxiety and test performance.

APPENDIX

Tables I–VI

* This table shows the square roots between the numbers 1 and 10. Others can be found by using the following information. Where n is any number between 1 and 10,

$\sqrt{n}$ can be found on table

$\sqrt{100n} = 10\sqrt{n}$

$\sqrt{10n}$ can be found on table

$\sqrt{\frac{1}{10}n} = \frac{1}{10}\sqrt{10n}$

Table I. Squares, Square Roots, and Reciprocals[*]

N	N^2	$\sqrt{N}$	$\sqrt{10N}$	N	N^2	$\sqrt{N}$	$\sqrt{10N}$
1.00	1.0000	1.00000	3.16228	1.50	2.2500	1.22474	3.87298
1.01	1.0201	1.00499	3.17805	1.51	2.2801	1.22882	3.88587
1.02	1.0404	1.00995	3.19374	1.52	2.3104	1.23288	3.89872
1 03	1.0609	1.01489	3.20936	1.53	2.3409	1.23693	3.91152
1.04	1.0816	1.01980	3.22490	1.54	2.3716	1.24097	3.92428
1.05	1.1025	1.02470	3.24037	1.55	2.4025	1.24499	3.93700
1.06	1.1236	1.02956	3.25576	1.56	2.4336	1.24900	3.94968
1.07	1.1449	1.03441	3.27109	1.57	2.4649	1.25300	3.96232
1.08	1.1664	1.03923	3.28634	1.58	2.4964	1.25698	3.97492
1.09	1.1881	1.04403	3.30151	1.59	2.5281	1.26095	3.98748
1.10	1.2100	1.04881	3.31662	1.60	2.5600	1.26491	4.00000
1.11	1.2321	1.05357	3.33167	1.61	2.5921	1.26886	4.01248
1.12	1.2544	1.05830	3.34664	1.62	2.6244	1.27279	4.02492
1.13	1.2769	1.06301	3.36155	1.63	2.6569	1.27671	4.03733
1.14	1.2996	1.06771	3.37639	1.64	2.6896	1.28062	4.04969
1.15	1.3225	1.07238	3.39116	1.65	2.7225	1.28452	4.06202
1.16	1.3456	1.07703	3.40588	1.66	2.7556	1.28841	4.07431
1.17	1.3689	1.08167	3.42053	1.67	2.7889	1.29228	4.08656
1.18	1.3924	1.08628	3.43511	1.68	2.8224	1.29615	4.09878
1.19	1.4161	1.09087	3.44964	1.69	2.8561	1.30000	4.11096
1.20	1.4400	1.09545	3.46410	1.70	2.8900	1.30384	4.12311
1.21	1.4641	1.10000	3.47851	1.71	2.9241	1.30767	4.13521
1.22	1.4884	1.10454	3.49285	1.72	2.9584	1.31149	4.14729
1.23	1.5129	1.10905	3.50714	1.73	2.9929	1.31529	4.15933
1.24	1.5376	1.11355	3.52136	1.74	3.0276	1.31909	4.17133
1.25	1.5625	1.11803	3.53553	1.75	3.0625	1.32288	4.18330
1.26	1.5876	1.12250	3.54965	1.76	3.0976	1.32665	4.19524
1.27	1.6129	1.12694	3.56371	1.77	3.1329	1.33041	4.20714
1.28	1.6384	1.13137	3.57771	1.78	3.1684	1.33417	4.21900
1.29	1.6641	1.13578	3.59166	1.79	3.2041	1.33791	4.23084
1.30	1.6900	1.14018	3.60555	1.80	3.2400	1.34164	4.24264
1.31	1.7161	1.14455	3.61939	1.81	3.2761	1.34536	4.25441
1.32	1.7424	1.14891	3.63318	1.82	3.3124	1.34907	4.26615
1.33	1.7689	1.15326	3.64692	1.83	3.3489	1.35277	4.27785
1.34	1.7956	1.15758	3.66060	1.84	3.3856	1.35647	4.28952
1.35	1.8225	1.16190	3.67423	1.85	3.4225	1.36015	4.30116
1.36	1.8496	1.16619	3.68782	1.86	3.4596	1.36382	4.31277
1.37	1.8769	1.17047	3.70135	1.87	3.4969	1.36748	4.32435
1.38	1.9044	1.17473	3.71484	1.88	3.5344	1.37113	4.33590
1.39	1.9321	1.17898	3.72827	1.89	3.5721	1.37477	4.34741
1.40	1.9600	1.18322	3.74166	1.90	3.6100	1.37840	4.35890
1.41	1.9881	1.18743	3.75500	1.91	3.6481	1.38203	4.37035
1.42	2.0164	1.19164	3.76829	1.92	3.6864	1.38564	4.38178
1.43	2.0449	1.19583	3.78153	1.93	3.7249	1.38924	4.39318
1.44	2.0736	1.20000	3.79473	1.94	3.7636	1.39284	4.40454
1.45	2.1025	1.20416	3.80789	1.95	3.8025	1.39642	4.41588
1.46	2.1316	1.20830	3.82099	1.96	3.8416	1.40000	4.42719
1.47	2.1609	1.21244	3.83406	1.97	3.8809	1.40357	4.43847
1.48	2.1904	1.21655	3.84708	1.98	3.9204	1.40712	4.44972
1.49	2.2201	1.22066	3.86005	1.99	3.9601	1.41067	4.46094
1.50	2.2500	1.22474	3.87298	2.00	4.0000	1.41421	4.47214
N	N^2	$\sqrt{N}$	$\sqrt{10N}$	N	N^2	$\sqrt{N}$	$\sqrt{10N}$

N	N^2	$\sqrt{N}$	$\sqrt{10N}$	N	N^2	$\sqrt{N}$	$\sqrt{10N}$
2.00	4.0000	1.41421	4.47214	**2.50**	6.25C0	1.58114	5.00000
2.01	4.0401	1.41774	4.48330	2.51	6.3001	1.58430	5.00999
2.02	4.0804	1.42127	4.49444	2.52	6.3504	1.58745	5.01996
2.03	4.1209	1.42478	4.50555	2.53	6.4009	1.59060	5.02991
2.04	4.1616	1.42829	4.51664	2.54	6.4516	1.59374	5.03984
2.05	4.2025	1.43178	4.52769	2.55	6.5025	1.59687	5.04975
2.06	4.2436	1.43527	4.53872	2.56	6.5536	1.60000	5.05964
2.07	4.2849	1.43875	4.54973	2.57	6.6049	1.60312	5.06952
2.08	4.3264	1.44222	4.56070	2.58	6.6564	1.60624	5.07937
2.09	4.3681	1.44568	4.57165	2.59	6.7081	1.60935	5.08920
2.10	4.4100	1.44914	4.58258	**2.60**	6.7600	1 61245	5.09902
2.11	4.4521	1.45258	4.59347	2.61	6.8121	1.61555	5.10882
2.12	4.4944	1.45602	4.60435	2.62	6.8644	1.61864	5.11859
2.13	4.5369	1.45945	4.61519	2.63	6.9169	1.62173	5.12835
2.14	4.5796	1.46287	4.62601	2.64	6.9696	1.62481	5.13809
2.15	4.6225	1.46629	4.63681	2.65	7.0225	1.62788	5.14782
2.16	4.6656	1.46969	4.64758	2.66	7.0756	1.63095	5.15752
2.17	4.7089	1.47309	4.65833	2.67	7.1289	1.63401	5.16720
2.18	4.7524	1.47648	4.66905	2.68	7.1824	1.63707	5.17687
2.19	4.7961	1.47986	4.67974	2.69	7.2361	1.64012	5.18652
2.20	4.8400	1.48324	4.69042	**2.70**	7.2900	1.64317	5.19615
2.21	4.8841	1.48661	4.70106	2.71	7.3441	1.64621	5.20577
2.22	4.9284	1.48997	4.71169	2.72	7.3984	1.64924	5.21536
2.23	4.9729	1.49332	4.72229	2.73	7.4529	1.65227	5.22494
2.24	5.0176	1.49666	4.73286	2.74	7.5076	1.65529	5.23450
2.25	5.0625	1.50000	4.74342	2.75	7.5625	1.65831	5.24404
2.26	5.1076	1.50333	4.75395	2.76	7.6176	1.66132	5.25357
2.27	5.1529	1.50665	4.76445	2.77	7.6729	1.66433	5.26308
2.28	5.1984	1.50997	4.77493	2.78	7.7284	1.66733	5.27257
2.29	5.2441	1.51327	4.78539	2.79	7.7841	1.67033	5.28205
2.30	5.2900	1.51658	4.79583	**2.80**	7.8400	1.67332	5.29150
2.31	5.3361	1.51987	4.80625	2.81	7.8961	1.67631	5.30094
2.32	5.3824	1.52315	4.81664	2.82	7.9524	1.67929	5.31037
2.33	5.4289	1.52643	4.82701	2.83	8.0089	1.68226	5.31977
2.34	5.4756	1.52971	4.83735	2.84	8.0656	1.68523	5.32917
2.35	5.5225	1.53297	4.84768	2.85	8.1225	1.68819	5.33854
2.36	5.5696	1.53623	4.85798	2.86	8.1796	1.69115	5.34790
2.37	5.6169	1.53948	4.86826	2.87	8.2369	1.69411	5.35724
2.38	5.6644	1.54272	4.87852	2.88	8.2944	1.69706	5.36656
2.39	5.7121	1.54596	4.88876	2.89	8.3521	1.70000	5.37587
2.40	5.7600	1.54919	4.89898	**2.90**	8.4100	1.70294	5.38516
2.41	5.8081	1.55242	4.90918	2.91	8.4681	1.70587	5.39444
2.42	5.8564	1.55563	4.91935	2.92	8.5264	1.70880	5.40370
2.43	5.9049	1.55885	4.92950	2.93	8.5849	1.71172	5.41295
2.44	5.9536	1.56205	4.93964	2.94	8.6436	1.71464	5.42218
2.45	6.0025	1.56525	4.94975	2.95	8.7025	1.71756	5.43139
2.46	6.0516	1.56844	4.95984	2.96	8.7616	1.72047	5.44059
2.47	6.1009	1.57162	4.96991	2.97	8.8209	1.72337	5.44977
2.48	6.1504	1.57480	4.97996	2.98	8.8804	1.72627	5.45894
2.49	6.2001	1.57797	4.98999	2.99	8.9401	1.72916	5.46809
2.50	6.2500	1.58114	5.00000	**3.00**	9.0000	1.73205	5.47723
N	N^2	$\sqrt{N}$	$\sqrt{10N}$	N	N^2	$\sqrt{N}$	$\sqrt{10N}$

N	N^2	$\sqrt{N}$	$\sqrt{10N}$	N	N^2	$\sqrt{N}$	$\sqrt{10N}$
3.00	9.0000	1.73205	5.47723	**3.50**	12.2500	1.87083	5.91608
3.01	9.0601	1.73494	5.48635	3.51	12.3201	1.87350	5.92453
3.02	9.1204	1.73781	5.49545	3.52	12.3904	1.87617	5.93296
3.03	9.1809	1.74069	5.50454	3.53	12.4609	1.87883	5.94138
3.04	9.2416	1.74356	5.51362	3.54	12.5316	1.88149	5.94979
3.05	9.3025	1.74642	5.52268	3.55	12.6025	1.88414	5.95819
3.06	9.3636	1.74929	5.53173	3.56	12.6736	1.88680	5.96657
3.07	9.4249	1.75214	5.54076	3.57	12.7449	1.88944	5.97495
3.08	9.4864	1.75499	5.54977	3.58	12.8164	1.89209	5.98331
3.09	9.5481	1.75784	5.55878	3.59	12.8881	1.89473	5.99166
3.10	9.6100	1.76068	5.56776	**3.60**	12.9600	1.89737	6.00000
3.11	9.6721	1.76352	5.57674	3.61	13.0321	1.90000	6.00833
3.12	9.7344	1.76635	5.58570	3.62	13.1044	1.90263	6.01664
3.13	9.7969	1.76918	5.59464	3.63	13.1769	1.90526	6.02495
3.14	9.8596	1.77200	5.60357	3.64	13.2496	1.90788	6.03324
3.15	9.9225	1.77482	5.61249	3.65	13.3225	1.91050	6.04152
3.16	9.9856	1.77764	5.62139	3.66	13.3956	1.91311	6.04979
3.17	10.0489	1.78045	5.63028	3.67	13.4689	1.91572	6.05805
3.18	10.1124	1.78326	5.63915	3.68	13.5424	1.91833	6.06630
3.19	10.1761	1.78606	5.64801	3.69	13.6161	1.92094	6.07454
3.20	10.2400	1.78885	5.65685	**3.70**	13.6900	1.92354	6.08276
3.21	10.3041	1.79165	5.66569	3.71	13.7641	1.92614	6.09098
3.22	10.3684	1.79444	5.67450	3.72	13.8384	1.92873	6.09918
3.23	10.4329	1.79722	5.68331	3.73	13.9129	1.93132	6.10737
3.24	10.4976	1.80000	5.69210	3.74	13.9876	1.93391	6.11555
3.25	10.5625	1.80278	5.70088	3.75	14.0625	1.93649	6.12372
3.26	10.6276	1.80555	5.70964	3.76	14.1376	1.93907	6.13188
3.27	10.6929	1.80831	5.71839	3.77	14.2129	1.94165	6.14003
3.28	10.7584	1.81108	5.72713	3.78	14.2884	1.94422	6.14817
3.29	10.8241	1.81384	5.73585	3.79	14.3641	1.94679	6.15630
3.30	10.8900	1.81659	5.74456	**3.80**	14.4400	1.94936	6.16441
3.31	10.9561	1.81934	5.75326	3.81	14.5161	1.95192	6.17252
3.32	11.0224	1.82209	5.76194	3.82	14.5924	1.95448	6.18061
3.33	11.0889	1.82483	5.77062	3.83	14.6689	1.95704	6.18870
3.34	11.1556	1.82757	5.77927	3.84	14.7456	1.95959	6.19677
3.35	11.2225	1.83030	5.78792	3.85	14.8225	1.96214	6.20484
3.36	11.2896	1.83303	5.79655	3.86	14.8996	1.96469	6.21289
3.37	11.3569	1.83576	5.80517	3.87	14.9769	1.96723	6.22093
3.38	11.4244	1.83848	5.81378	3.88	15.0544	1.96977	6.22896
3.39	11.4921	1.84120	5.82237	3.89	15.1321	1.97231	6.23699
3.40	11.5600	1.84391	5.83095	**3.90**	15.2100	1.97484	6.24500
3.41'	11.6281	1.84662	5.83952	3.91	15.2881	1.97737	6.25300
3.42	11.6964	1.84932	5.84808	3.92	15.3664	1.97990	6.26099
3.43	11.7649	1.85203	5.85662	3.93	15.4449	1.98242	6.26897
3.44	11.8336	1.85472	5.86515	3.94	15.5236	1.98494	6.27694
3.45	11.9025	1.85742	5.87367	3.95	15.6025	1.98746	6.28490
3.46	11.9716	1.86011	5.88218	3.96	15.6816	1.98997	6.29285
3.47	12.0409	1.86279	5.89067	3.97	15.7609	1.99249	6.30079
3.48	12.1104	1.86548	5.89915	3.98	15.8404	1.99499	6.30872
3.49	12.1801	1.86815	5.90762	3.99	15.9201	1.99750	6.31664
3.50	12.2500	1.87083	5.91608	**4.00**	16.0000	2.00000	6.32456
N	N^2	$\sqrt{N}$	$\sqrt{10N}$	N	N^2	$\sqrt{N}$	$\sqrt{10N}$

N	N²	√N	√10N	N	N²	√N	√10N
4.00	16.0000	2.00000	6.32456	**4.50**	20.2500	2.12132	6.70820
4.01	16.0801	2.00250	6.33246	4.51	20.3401	2.12368	6.71565
4.02	16.1604	2.00499	6.34035	4.52	20.4304	2.12603	6.72309
4.03	16.2409	2.00749	6.34823	4.53	20.5209	2.12838	6.73053
4.04	16.3216	2.00998	6.35610	4.54	20.6116	2.13073	6.73795
4.05	16.4025	2.01246	6.36396	4.55	20.7025	2.13307	6.74537
4.06	16.4836	2.01494	6.37181	4.56	20.7936	2.13542	6.75278
4.07	16.5649	2.01742	6.37966	4.57	20.8849	2.13776	6.76018
4.08	16.6464	2.01990	6.38749	4.58	20.9764	2.14009	6.76757
4.09	16.7281	2.02237	6.39531	4.59	21.0681	2.14243	6.77495
4.10	16.8100	2.02485	6.40312	**4.60**	21.1600	2.14476	6.78233
4.11	16.8921	2.02731	6.41093	4.61	21.2521	2.14709	6.78970
4.12	16.9744	2.02978	6.41872	4.62	21.3444	2.14942	6.79706
4.13	17.0569	2.03224	6.42651	4.63	21.4369	2.15174	6.80441
4.14	17.1396	2.03470	6.43428	4.64	21.5296	2.15407	6.81175
4.15	17.2225	2.03715	6.44205	4.65	21.6225	2.15639	6.81909
4.16	17.3056	2.03961	6.44981	4.66	21.7156	2.15870	6.82642
4.17	17.3889	2.04206	6.45755	4.67	21.8089	2.16102	6.83374
4.18	17.4724	2.04450	6.46529	4.68	21.9024	2.16333	6.84105
4.19	17.5561	2.04695	6.47302	4.69	21.9961	2.16564	6.84836
4.20	17.6400	2.04939	6.48074	**4.70**	22.0900	2.16795	6.85565
4.21	17.7241	2.05183	6.48845	4.71	22.1841	2.17025	6.86294
4.22	17.8084	2.05426	6.49615	4.72	22.2784	2.17256	6.87023
4.23	17.8929	2.05670	6.50384	4.73	22.3729	2.17486	6.87750
4.24	17.9776	2.05913	6.51153	4.74	22.4676	2.17715	6.88477
4.25	13.0625	2.06155	6.51920	4.75	22.5625	2.17945	6.89202
4.26	18.1476	2.06398	6.52687	4.76	22.6576	2.18174	6.89928
4.27	18.2329	2.06640	6.53452	4.77	22.7529	2.18403	6.90652
4.28	18.3184	2.06882	6.54217	4.78	22.8484	2.18632	6.91375
4.29	18.4041	2.07123	6.54981	4.79	22.9441	2.18861	6.92098
4.30	18.4900	2.07364	6.55744	**4.80**	23.0400	2.19089	6.92820
4.31	18.5761	2.07605	6.56506	4.81	23.1361	2.19317	6.93542
4.32	18.6624	2.07846	6.57267	4.82	23.2324	2.19545	6.94262
4.33	18.7489	2.08087	6.58027	4.83	23.3289	2.19773	6.94982
4.34	18.8356	2.08327	6.58787	4.84	23.4256	2.20000	6.95701
4.35	18.9225	2.08567	6.59545	4.85	23.5225	2.20227	6.96419
4.36	19.0096	2.08806	6.60303	4.86	23.6196	2.20454	6.97137
4.37	19.0969	2.09045	6.61060	4.87	23.7169	2.20681	6.97854
4.38	19.1844	2.09284	6.61816	4.88	23.8144	2.20907	6.98570
4.39	19.2721	2.09523	6.62571	4.89	23.9121	2.21133	6.99285
4.40	19.3600	2.09762	6.63325	**4.90**	24.0100	2.21359	7.00000
4.41	19.4481	2.10000	6.64078	4.91	24.1081	2.21585	7.00714
4.42	19.5364	2.10238	6.64831	4.92	24.2064	2.21811	7.01427
4.43	19.6249	2.10476	6.65582	4.93	24.3049	2.22036	7.02140
4.44	19.7136	2.10713	6.66333	4.94	24.4036	2.22261	7.02851
4.45	19.8025	2.10950	6.67083	4.95	24.5025	2.22486	7.03562
4.46	19.8916	2.11187	6.67832	4.96	24.6016	2.22711	7.04273
4.47	19.9809	2.11424	6.68581	4.97	24.7009	2.22935	7.04982
4.48	20.0704	2.11660	6.69328	4.98	24.8004	2.23159	7.05691
4.49	20.1601	2.11896	6.70075	4.99	24.9001	2.23383	7.06399
4.50	20.2500	2.12132	6.70820	**5.00**	25.0000	2.23607	7.07107
N	N²	√N	√10N	N	N²	√N	√10N

N	N^2	$\sqrt{N}$	$\sqrt{10N}$	N	N^2	$\sqrt{N}$	$\sqrt{10N}$
5.00	25.0000	2.23607	7.07107	**5.50**	30.2500	2.34521	7.41620
5.01	25.1001	2.23830	7.07814	5.51	30.3601	2.34734	7.42294
5.02	25.2004	2.24054	7.08520	5.52	30.4704	2.34947	7.42967
5.03	25.3009	2.24277	7.09225	5.53	30.5809	2.35160	7.43640
5.04	25.4016	2.24499	7.09930	5.54	30.6916	2.35372	7.44312
5.05	25.5025	2.24722	7.10634	5.55	30.8025	2.35584	7.44983
5.06	25.6036	2.24944	7.11337	5.56	30.9136	2.35797	7.45654
5.07	25.7049	2.25167	7.12039	5.57	31.0249	2.36008	7.46324
5.08	25.8064	2.25389	7.12741	5.58	31.1364	2.36220	7.46994
5.09	25.9081	2.25610	7.13442	5.59	31.2481	2.36432	7.47663
5.10	26.0100	2.25832	7.14143	**5.60**	31.3600	2.36643	7.48331
5.11	26.1121	2.26053	7.14843	5.61	31.4721	2.36854	7.48999
5.12	26.2144	2.26274	7.15542	5.62	31.5844	2.37065	7.49667
5.13	26.3169	2.26495	7.16240	5.63	31.6969	2.37276	7.50333
5.14	26.4196	2.26716	7.16938	5.64	31.8096	2.37487	7.50999
5.15	26.5225	2.26936	7.17635	5.65	31.9225	2.37697	7.51665
5.16	26.6256	2.27156	7.18331	5.66	32.0356	2.37908	7.52330
5.17	26.7289	2.27376	7.19027	5.67	32.1489	2.38118	7.52994
5.18	26.8324	2.27596	7.19722	5.68	32.2624	2.38328	7.53658
5.19	26.9361	2.27816	7.20417	5.69	32.3761	2.38537	7.54321
5.20	27.0400	2.28035	7.21110	**5.70**	32.4900	2.38747	7.54983
5.21	27.1441	2.28254	7.21803	5.71	32.6041	2.38956	7.55645
5.22	27.2484	2.28473	7.22496	5.72	32.7184	2.39165	7.56307
5.23	27.3529	2.28692	7.23187	5.73	32.8329	2.39374	7.56968
5.24	27.4576	2.28910	7.23878	5.74	32.9476	2.39583	7.57628
5.25	27.5625	2.29129	7.24569	5.75	33.0625	2.39792	7.58288
5.26	27.6676	2.29347	7.25259	5.76	33.1776	2.40000	7.58947
5.27	27.7729	2.29565	7.25948	5.77	33.2929	2.40208	7.59605
5.28	27.8784	2.29783	7.26636	5.78	33.4084	2.40416	7.60263
5.29	27.9841	2.30000	7.27324	5.79	33.5241	2.40624	7.60920
5.30	28.0900	2.30217	7.28011	**5.80**	33.6400	2.40832	7.61577
5.31	28.1961	2.30434	7.28697	5.81	33.7561	2.41039	7.62234
5.32	28.3024	2.30651	7.29383	5.82	33.8724	2.41247	7.62889
5.33	28.4089	2.30868	7.30068	5.83	33.9889	2.41454	7.63544
5.34	28.5156	2.31084	7.30753	5.84	34.1056	2.41661	7.64199
5.35	28.6225	2.31301	7.31437	5.85	34.2225	2.41868	7.64853
5.36	28.7296	2.31517	7.32120	5.86	34.3396	2.42074	7.65506
5.37	28.8369	2.31733	7.32803	5.87	34.4569	2.42281	7.66159
5.38	28.9444	2.31948	7.33485	5.88	34.5744	2.42487	7.66812
5.39	29.0521	2.32164	7.34166	5.89	34.6921	2.42693	7.67463
5.40	29.1600	2.32379	7.34847	**5.90**	34.8100	2.42899	7.68115
5.41	29.2681	2.32594	7.35527	5.91	34.9281	2.43105	7.68765
5.42	29.3764	2.32809	7.36206	5.92	35.0464	2.43311	7.69415
5.43	29.4849	2.33024	7.36885	5.93	35.1649	2.43516	7.70065
5.44	29.5936	2.33238	7.37564	5.94	35.2836	2.43721	7.70714
5.45	29.7025	2.33452	7.38241	5.95	35.4025	2.43926	7.71362
5.46	29.8116	2.33666	7.38918	5.96	35.5216	2.44131	7.72010
5.47	29.9209	2.33880	7.39594	5.97	35.6409	2.44336	7.72658
5.48	30.0304	2.34094	7.40270	5.98	35.7604	2.44540	7.73305
5.49	30.1401	2.34307	7.40945	5.99	35.8801	2.44745	7.73951
5.50	30.2500	2.34521	7.41620	**6.00**	36.0000	2.44949	7.74597
N	N^2	$\sqrt{N}$	$\sqrt{10N}$	N	N^2	$\sqrt{N}$	$\sqrt{10N}$

N	N^2	$\sqrt{N}$	$\sqrt{10N}$	N	N^2	$\sqrt{N}$	$\sqrt{10N}$
6.00	36.0000	2.44949	7.74597	**6.50**	42.2500	2.54951	8.06226
6.01	36.1201	2.45153	7.75242	6.51	42.3801	2.55147	8.06846
6.02	36.2404	2.45357	7.75887	6.52	42.5104	2.55343	8.07465
6.03	36.3609	2.45561	7.76531	6.53	42.6409	2.55539	8.08084
6.04	36.4816	2.45764	7.77174	6.54	42.7716	2.55734	8.08703
6.05	36.6025	2.45967	7.77817	6.55	42.9025	2.55930	8.09321
6.06	36.7236	2.46171	7.78460	6.56	43.0336	2.56125	8.09938
6.07	36.8449	2.46374	7.79102	6.57	43.1649	2.56320	8.10555
6.08	36.9664	2.46577	7.79744	6.58	43.2964	2.56515	8.11172
6.09	37.0881	2.46779	7.80385	6.59	43.4281	2.56710	8.11788
6.10	37.2100	2.46982	7.81025	**6.60**	43.5600	2.56905	8.12404
6.11	37.3321	2.47184	7.81665	6.61	43.6921	2.57099	8.13019
6.12	37.4544	2.47386	7.82304	6.62	43.8244	2.57294	8.13634
6.13	37.5769	2.47588	7.82943	6.63	43.9569	2.57488	8.14248
6.14	37.6996	2.47790	7.83582	6.64	44.0896	2.57682	8.14862
6.15	37.8225	2.47992	7.84219	6.65	44.2225	2.57876	8.15475
6.16	37.9456	2.48193	7.84857	6.66	44.3556	2.58070	8.16088
6.17	38.0689	2.48395	7.85493	6.67	44.4889	2.58263	8.16701
6.18	38.1924	2.48596	7.86130	6.68	44.6224	2.58457	8.17313
6.19	38.3161	2.48797	7.86766	6.69	44.7561	2.58650	8.17924
6.20	38.4400	2.48998	7.87401	**6.70**	44.8900	2.58844	8.18535
6.21	38.5641	2.49199	7.88036	6.71	45.0241	2.59037	8.19146
6.22	38.6884	2.49399	7.88670	6.72	45.1584	2.59230	8.19756
6.23	38.8129	2.49600	7.89303	6.73	45.2929	2.59422	8.20366
6.24	38.9376	2.49800	7.89937	6.74	45.4276	2.59615	8.20975
6.25	39.0625	2.50000	7.90569	6.75	45.5625	2.59808	8.21584
6.26	39.1876	2.50200	7.91202	6.76	45.6976	2.60000	8.22192
6.27	39.3129	2.50400	7.91833	6.77	45.8329	2.60192	8.22800
6.28	39.4384	2.50599	7.92465	6.78	45.9684	2.60384	8.23408
6.29	39.5641	2.50799	7.93095	6.79	46.1041	2.60576	8.24015
6.30	39.6900	2.50998	7.93725	**6.80**	46.2400	2.60768	8.24621
6.31	39.8161	2.51197	7.94355	6.81	46.3761	2.60960	8.25227
6.32	39.9424	2.51396	7.94984	6.82	46.5124	2.61151	8.25833
6.33	40.0689	2.51595	7.95613	6.83	46.6489	2.61343	8.26438
6.34	40.1956	2.51794	7.96241	6.84	46.7856	2.61534	8.27043
6.35	40.3225	2.51992	7.96869	6.85	46.9225	2.61725	8.27647
6.36	40.4496	2.52190	7.97496	6.86	47.0596	2.61916	8.28251
6.37	40.5769	2.52389	7.98123	6.87	47.1969	2.62107	8.28855
6.38	40.7044	2.52587	7.98749	6.88	47.3344	2.62298	8.29458
6.39	40.8321	2.52784	7.99375	6.89	47.4721	2.62488	8.30060
6.40	40.9600	2.52982	8.00000	**6.90**	47.6100	2.62679	8.30662
6.41	41.0881	2.53180	8.00625	6.91	47.7481	2.62869	8.31264
6.42	41.2164	2.53377	8.01249	6.92	47.8864	2.63059	8.31865
6.43	41.3449	2.53574	8.01873	6.93	48.0249	2.63249	8.32466
6.44	41.4736	2.53772	8.02496	6.94	48.1636	2.63439	8.33067
6.45	41.6025	2.53969	8.03119	6.95	48.3025	2.63629	8.33667
6.46	41.7316	2.54165	8.03741	6.96	48.4416	2.63818	8.34266
6.47	41.8609	2.54362	8.04363	6.97	48.5809	2.64008	8.34865
6.48	41.9904	2.54558	8.04984	6.98	48.7204	2.64197	8.35464
6.49	42.1201	2.54755	8.05605	6.99	48.8601	2.64386	8.36062
6.50	42.2500	2.54951	8.06226	**7.00**	49.0000	2.64575	8.36660
N	N^2	$\sqrt{N}$	$\sqrt{10N}$	N	N^2	$\sqrt{N}$	$\sqrt{10N}$

N	N^2	$\sqrt{N}$	$\sqrt{10N}$	N	N^2	$\sqrt{N}$	$\sqrt{10N}$
7.00	49.0000	2.64575	8.36660	**7.50**	56.2500	2.73861	8.66025
7.01	49.1401	2.64764	8.37257	7.51	56.4001	2.74044	8.66603
7.02	49.2804	2.64953	8.37854	7.52	56.5504	2.74226	8.67179
7.03	49.4209	2.65141	8.38451	7.53	56.7009	2.74408	8.67756
7.04	49.5616	2.65330	8.39047	7.54	56.8516	2.74591	8.68332
7.05	49.7025	2.65518	8.39643	7.55	57.0025	2.74773	8.68907
7.06	49.8436	2.65707	8.40238	7.56	57.1536	2.74955	8.69483
7.07	49.9849	2.65895	8.40833	7.57	57.3049	2.75136	8.70057
7.08	50.1264	2.66083	8.41427	7.58	57.4564	2.75318	8.70632
7.09	50.2681	2.66271	8.42021	7.59	57.6081	2.75500	8.71206
7.10	50.4100	2.66458	8.42615	**7.60**	57.7600	2.75681	8.71780
7.11	50.5521	2.66646	8.43208	7.61	57.9121	2.75862	8.72353
7.12	50.6944	2.66833	8.43801	7.62	58.0644	2.76043	8.72926
7.13	50.8369	2.67021	8.44393	7.63	58.2169	2.76225	8.73499
7.14	50.9796	2.67208	8.44985	7.64	58.3696	2.76405	8.74071
7.15	51.1225	2.67395	8.45577	7.65	58.5225	2.76586	8.74643
7.16	51.2656	2.67582	8.46168	7.66	58.6756	2.76767	8.75214
7.17	51.4089	2.67769	8.46759	7.67	58.8289	2.76948	8.75785
7.18	51.5524	2.67955	8.47349	7.68	58.9824	2.77128	8.76356
7.19	51.6961	2.68142	8.47939	7.69	59.1361	2.77308	8.76926
7.20	51.8400	2.68328	8.48528	**7.70**	59.2900	2.77489	8.77496
7.21	51.9841	2.68514	8.49117	7.71	59.4441	2.77669	8.78066
7.22	52.1284	2.68701	8.49706	7.72	59.5984	2.77849	8.78635
7.23	52.2729	2.68887	8.50294	7.73	59.7529	2.78029	8.79204
7.24	52.4176	2.69072	8.50882	7.74	59.9076	2.78209	8.79773
7.25	52.5625	2.69258	8.51469	7.75	60.0625	2.78388	8.80341
7.26	52.7076	2.69444	8.52056	7.76	60.2176	2.78568	8.80909
7.27	52.8529	2.69629	8.52643	7.77	60.3729	2.78747	8.81476
7.28	52.9984	2.69815	8.53229	7.78	60.5284	2.78927	8.82043
7.29	53.1441	2.70000	8.53815	7.79	60.6841	2.79106	8.82610
7.30	53.2900	2.70185	8.54400	**7.80**	60.8400	2.79285	8.83176
7.31	53.4361	2.70370	8.54985	7.81	60.9961	2.79464	8.83742
7.32	53.5824	2.70555	8.55570	7.82	61.1524	2.79643	8.84308
7.33	53.7289	2.70740	8.56154	7.83	61.3089	2.79821	8.84873
7.34	53.8756	2.70924	8.56738	7.84	61.4656	2.80000	8.85438
7.35	54.0225	2.71109	8.57321	7.85	61.6225	2.80179	8.86002
7.36	54.1696	2.71293	8.57904	7.86	61.7796	2.80357	8.86566
7.37	54.3169	2.71477	8.58487	7.87	61.9369	2.80535	8.87130
7.38	54.4644	2.71662	8.59069	7.88	62.0944	2.80713	8.87694
7.39	54.6121	2.71846	8.59651	7.89	62.2521	2.80891	8.88257
7.40	54.7600	2.72029	8.60233	**7.90**	62.4100	2.81069	8.88819
7.41	54.9081	2.72213	8.60814	7.91	62.5681	2.81247	8.89382
7.42	55.0564	2.72397	8.61394	7.92	62.7264	2.81425	8.89944
7.43	55.2049	2.72580	8.61974	7.93	62.8849	2.81603	8.90505
7.44	55.3536	2.72764	8.62554	7.94	63.0436	2.81780	8.91067
7.45	55.5025	2.72947	8.63134	7.95	63.2025	2.81957	8.91628
7.46	55.6516	2.73130	8.63713	7.96	63.3616	2.82135	8.92188
7.47	55.8009	2.73313	8.64292	7.97	63.5209	2.82312	8.92749
7.48	55.9504	2.73496	8.64870	7.98	63.6804	2.82489	8.93308
7.49	56.1001	2.73679	8.65448	7.99	63.8401	2.82666	8.93868
7.50	56.2500	2.73861	8.66025	**8.00**	64.0000	2.82843	8.94427
N	N^2	$\sqrt{N}$	$\sqrt{10N}$	N	N^2	$\sqrt{N}$	$\sqrt{10N}$

N	N^2	$\sqrt{N}$	$\sqrt{10N}$	N	N^2	$\sqrt{N}$	$\sqrt{10N}$
8.00	64.0000	2.82843	8.94427	**8.50**	72.2500	2.91548	9.21954
8.01	64.1601	2.83019	8.94986	8.51	72.4201	2.91719	9.22497
8.02	64.3204	2.83196	8.95545	8.52	72.5904	2.91890	9.23038
8.03	64.4809	2.83373	8.96103	8.53	72.7609	2.92062	9.23580
8.04	64.6416	2.83549	8.96660	8.54	72.9316	2.92233	9.24121
8.05	64.8025	2.83725	8.97218	8.55	73.1025	2.92404	9.24662
8.06	64.9636	2.83901	8.97775	8.56	73.2736	2.92575	9.25203
•8.07	65.1249	2.84077	8.98332	8.57	73.4449	2.92746	9.25743
8.08	65.2864	2.84253	8.98888	8.58	73.6164	2.92916	9.26283
8.09	65.4481	2.84429	8.99444	8.59	73.7881	2.93087	9.26823
8.10	65.6100	2.84605	9.00000	**8.60**	73.9600	2.93258	9.27362
8.11	65.7721	2.84781	9.00555	8.61	74.1321	2.93428	9.27901
8.12	65.9344	2.84956	9.01110	8.62	74.3044	2.93598	9.28440
8.13	66.0969	2.85132	9.01665	8.63	74.4769	2.93769	9.28978
8.14	66.2596	2.85307	9.02219	8.64	74.6496	2.93939	9.29516
8.15	66.4225	2.85482	9.02774	8.65	74.8225	2.94109	9.30054
8.16	66.5856	2.85657	9.03327	8.66	74.9956	2.94279	9.30591
8.17	66.7489	2.85832	9.03881	8.67	75.1689	2.94449	9.31128
8.18	66.9124	2.86007	9.04434	8.68	75.3424	2.94618	9.31665
8.19	67.0761	2.86182	9.04986	8.69	75.5161	2.94788	9.32202
8.20	67.2400	2.86356	9.05539	**8.70**	75.6900	2.94958	9.32738
8.21	67.4041	2.86531	9.06091	8.71	75.8641	2.95127	9.33274
8.22	67.5684	2.86705	9.06642	8.72	76.0384	2.95296	9.33809
8.23	67.7329	2.86880	9.07193	8.73	76.2129	2.95466	9.34345
8.24	67.8976	2.87054	9.07744	8.74	76.3876	2.95635	9.34880
8.25	68.0625	2.87228	9.08295	8.75	76.5625	2.95804	9.35414
8.26	68.2276	2.87402	9.08845	8.76	76.7376	2.95973	9.35949
8.27	68.3929	2.87576	9.09395	8.77	76.9129	2.96142	9.36483
8.28	68.5584	2.87750	9.09945	8.78	77.0884	2.96311	9.37017
8.29	68.7241	2.87924	9.10494	8.79	77.2641	2.96469	9.37550
8.30	68.8900	2.88097	9.11043	**8.80**	77.4400	2.96648	9.38083
8.31	69.0561	2.88271	9.11592	8.81	77.6161	2.96816	9.38616
8.32	69.2224	2.88444	9.12140	8.82	77.7924	2.96985	9.39149
8.33	69.3889	2.88617	9.12688	8.83	77.9689	2.97153	9.39681
8.34	69.5556	2.88791	9.13236	8.84	78.1456	2.97321	9.40213
8.35	69.7225	2.88964	9.13783	8.85	78.3225	2.97489	9.40744
8.36	69.8896	2.89137	9.14330	8.86	78.4996	2.97658	9.41276
8.37	70.0569	2.89310	9.14877	8.87	78.6769	2.97825	9.41807
8.38	70.2244	2.89482	9.15423	8.88	78.8544	2.97993	9.42338
8.39	70.3921	2.89655	9.15969	8.89	79.0321	2.98161	9.42868
8.40	70.5600	2.89828	9.16515	**8.90**	79.2100	2.98329	9.43398
8.41	70.7281	2.90000	9.17061	8.91	79.3881	2.98496	9.43928
8.42	70.8964	2.90172	9.17606	8.92	79.5664	2.98664	9.44458
8.43	71.0649	2.90345	9.18150	8.93	79.7449	2.98831	9.44987
8.44	71.2336	2.90517	9.18695	8.94	79.9236	2.98998	9.45516
8.45	71.4025	2.90689	9.19239	8.95	80.1025	2.99166	9.46044
8.46	71.5716	2.90861	9.19783	8.96	80.2816	2.99333	9.46573
8.47	71.7409	2.91033	9.20326	8.97	80.4609	2.99500	9.47101
8.48	71.9104	2.91204	9.20869	8.98	80.6404	2.99666	9.47629
8.49	72.0801	2.91376	9.21412	8.99	80.8201	2.99833	9.48156
8.50	72.2500	2.91548	9.21954	**9.00**	81.0000	3.00000	9.48683
N	N^2	$\sqrt{N}$	$\sqrt{10N}$	N	N^2	$\sqrt{N}$	$\sqrt{10N}$

N	N^2	$\sqrt{N}$	$\sqrt{10N}$	N	N^2	$\sqrt{N}$	$\sqrt{10N}$
9.00	81.0000	3.00000	9.48683	**9.50**	90.2500	3.08221	9.74679
9.01	81.1801	3.00167	9.49210	9.51	90.4401	3.08383	9.75192
9.02	81.3604	3.00333	9.49737	9.52	90.6304	3.08545	9.75705
9.03	81.5409	3.00500	9.50263	9.53	90.8209	3.08707	9.76217
9.04	81.7216	3.00666	9.50789	9.54	91.0116	3.08869	9.76729
9.05	81.9025	3.00832	9.51315	9.55	91.2025	3.09031	9.77241
9.06	82.0836	3.00998	9.51840	9.56	91.3936	3.09192	9.77753
9.07	82.2649	3.01164	9.52365	9.57	91.5849	3.09354	9.78264
9.08	82.4464	3.01330	9.52890	9.58	91.7764	3.09516	9.78775
9.09	82.6281	3.01496	9.53415	9.59	91.9681	3.09677	9.79285
9.10	82.8100	3.01662	9.53939	**9.60**	92.1600	3.09839	9.79796
9.11	82.9921	3.01828	9.54463	9.61	92.3521	3.10000	9.80306
9.12	83.1744	3.01993	9.54987	9.62	92.5444	3.10161	9.80816
9.13	83.3569	3.02159	9.55510	9.63	92.7369	3.10322	9.81326
9.14	83.5396	3.02324	9.56033	9.64	92.9296	3.10483	9.81835
9.15	83.7225	3.02490	9.56556	9.65	93.1225	3.10644	9.82344
9.16	83.9056	3.02655	9.57079	9.66	93.3156	3.10805	9.82853
9.17	84.0889	3.02820	9.57601	9.67	93.5089	3.10966	9.83362
9.18	84.2724	3.02985	9.58123	9.68	93.7024	3.11127	9.83870
9.19	84.4561	3.03150	9.58645	9.69	93.8961	3.11288	9.84378
9.20	84.6400	3.03315	9.59166	**9.70**	94.0900	3.11448	9.84886
9.21	84.8241	3.03480	9.59687	9.71	94.2841	3.11609	9.85393
9.22	85.0084	3.03645	9.60208	9.72	94.4784	3.11769	9.85901
9.23	85.1929	3.03809	9.60729	9.73	94.6729	3.11929	9.86408
9.24	85.3776	3.03974	9.61249	9.74	94.8676	3.12090	9.86914
9.25	85.5625	3.04138	9.61769	9.75	95.0625	3.12250	9.87421
9.26	85.7476	3.04302	9.62289	9.76	95.2576	3.12410	9.87927
9.27	85.9329	3.04467	9.62808	9.77	95.4529	3.12570	9.88433
9.28	86.1184	3.04631	9.63328	9.78	95.6484	3.12730	9.88939
9.29	86.3041	3.04795	9.63846	9.79	95.8441	3.12890	9.89444
9.30	86.4900	3.04959	9.64365	**9.80**	96.0400	3.13050	9.89949
9.31	86.6761	3.05123	9.64883	9.81	96.2361	3.13209	9.90454
9.32	86.8624	3.05287	9.65401	9.82	96.4324	3.13369	9.90959
9.33	87.0489	3.05450	9.65919	9.83	96.6289	3.13528	9.91464
9.34	87.2356	3.05614	9.66437	9.84	96.8256	3.13688	9.91968
9.35	87.4225	3.05778	9.66954	9.85	97.0225	3.13847	9.92472
9.36	87.6096	3.05941	9.67471	9.86	97.2196	3.14006	9.92975
9.37	87.7969	3.06105	9.67988	9.87	97.4169	3.14166	9.93479
9.38	87.9844	3.06268	9.68504	9.88	97.6144	3.14325	9.93982
9.39	88.1721	3.06431	9.69020	9.89	97.8121	3.14484	9.94485
9.40	88.3600	3.06594	9.69536	**9.90**	98.0100	3.14643	9.94987
9.41	88.5481	3.06757	9.70052	9.91	98.2081	3.14802	9.95490
9.42	88.7364	3.06920	9.70567	9.92	98.4064	3.14960	9.95992
9.43	88.9249	3.07083	9.71082	9.93	98.6049	3.15119	9.96494
9.44	89.1136	3.07246	9.71597	9.94	98.8036	3.15278	9.96995
9.45	89.3025	3.07409	9.72111	9.95	99.0025	3.15436	9.97497
9.46	89.4916	3.07571	9.72625	9.96	99.2016	3.15595	9.97998
9.47	89.6809	3.07734	9.73139	9.97	99.4009	3.15753	9.98499
9.48	89.8704	3.07896	9.73653	9.98	99.6004	3.15911	9.98999
9.49	90.0601	3.08058	9.74166	9.99	99.8001	3.16070	9.99500
9.50	90.2500	3.08221	9.74679	**10.00**	100.000	3.16228	10.0000
N	N^2	$\sqrt{N}$	$\sqrt{10N}$	N	N^2	$\sqrt{N}$	$\sqrt{10N}$

Table II. The Unit Normal Distribution

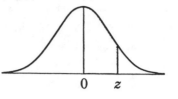

AREAS
under the
STANDARD
NORMAL CURVE
from 0 to z

z	0	1	2	3	4	5	6	7	8	9
0.0	.0000	.0040	.0080	.0120	.0160	.0199	.0239	.0279	.0319	.0359
0.1	.0398	.0438	.0478	.0517	.0557	.0596	.0636	.0675	.0714	.0754
0.2	.0793	.0832	.0871	.0910	.0948	.0987	.1026	.1064	.1103	.1141
0.3	.1179	.1217	.1255	.1293	.1331	.1368	.1406	.1443	.1480	.1517
0.4	.1554	.1591	.1628	.1664	.1700	.1736	.1772	.1808	.1844	.1879
0.5	.1915	.1950	.1985	.2019	.2054	.2088	.2123	.2157	.2190	.2224
0.6	.2258	.2291	.2324	.2357	.2389	.2422	.2454	.2486	.2518	.2549
0.7	.2580	.2612	.2642	.2673	.2704	.2734	.2764	.2794	.2823	.2852
0.8	.2881	.2910	.2939	.2967	.2996	.3023	.3051	.3078	.3106	.3133
0.9	.3159	.3186	.3212	.3238	.3264	.3289	.3315	.3340	.3365	.3389
1.0	.3413	.3438	.3461	.3485	.3508	.3531	.3554	.3577	.3599	.3621
1.1	.3643	.3665	.3686	.3708	.3729	.3749	.3770	.3790	.3810	.3830
1.2	.3849	.3869	.3888	.3907	.3925	.3944	.3962	.3980	.3997	.4015
1.3	.4032	.4049	.4066	.4082	.4099	.4115	.4131	.4147	.4162	.4177
1.4	.4192	.4207	.4222	.4236	.4251	.4265	.4279	.4292	.4306	.4319
1.5	.4332	.4345	.4357	.4370	.4382	.4394	.4406	.4418	.4429	.4441
1.6	.4452	.4463	.4474	.4484	.4495	.4505	.4515	.4525	.4535	.4545
1.7	.4554	.4564	.4573	.4582	.4591	.4599	.4608	.4616	.4625	.4633
1.8	.4641	.4649	.4656	.4664	.4671	.4678	.4686	.4693	.4699	.4706
1.9	.4713	.4719	.4726	.4732	.4738	.4744	.4750	.4756	.4761	.4767
2.0	.4772	.4778	.4783	.4788	.4793	.4798	.4803	.4808	.4812	.4817
2.1	.4821	.4826	.4830	.4834	.4838	.4842	.4846	.4850	.4854	.4857
2.2	.4861	.4864	.4868	.4871	.4875	.4878	.4881	.4884	.4887	.4890
2.3	.4893	.4896	.4898	.4901	.4904	.4906	.4909	.4911	.4913	.4916
2.4	.4918	.4920	.4922	.4925	.4927	.4929	.4931	.4932	.4934	.4936
2.5	.4938	.4940	.4941	.4943	.4945	.4946	.4948	.4949	.4951	.4952
2.6	.4953	.4955	.4956	.4957	.4959	.4960	.4961	.4962	.4963	.4964
2.7	.4965	.4966	.4967	.4968	.4969	.4970	.4971	.4972	.4973	.4974
2.8	.4974	.4975	.4976	.4977	.4977	.4978	.4979	.4979	.4980	.4981
2.9	.4981	.4982	.4982	.4983	.4984	.4984	.4985	.4985	.4986	.4986
3.0	.4987	.4987	.4987	.4988	.4988	.4989	.4989	.4989	.4990	.4990
3.1	.4990	.4991	.4991	.4991	.4992	.4992	.4992	.4992	.4993	.4993
3.2	.4993	.4993	.4994	.4994	.4994	.4994	.4994	.4995	.4995	.4995
3.3	.4995	.4995	.4995	.4996	.4996	.4996	.4996	.4996	.4996	.4997
3.4	.4997	.4997	.4997	.4997	.4997	.4997	.4997	.4997	.4997	.4998
3.5	.4998	.4998	.4998	.4998	.4998	.4998	.4998	.4998	.4998	.4998
3.6	.4998	.4998	.4999	.4999	.4999	.4999	.4999	.4999	.4999	.4999
3.7	.4999	.4999	.4999	.4999	.4999	.4999	.4999	.4999	.4999	.4999
3.8	.4999	.4999	.4999	.4999	.4999	.4999	.4999	.4999	.4999	.4999
3.9	.5000	.5000	.5000	.5000	.5000	.5000	.5000	.5000	.5000	.5000

Table III. Random Numbers

51772	74640	42331	29044	46621
24033	23491	83587	06568	21960
45939	60173	52078	25424	11645
30586	02133	75797	45406	31041
03585	79353	81938	82322	96799
64937	03355	95863	20790	65304
15630	64759	51135	98527	62586
09448	56301	57683	30277	94623
21631	91157	77331	60710	52290
91097	17480	29414	06829	87843
50532	25496	95652	42457	73547
07136	40876	79971	54195	25708
27989	64728	10744	08396	56242
85184	73949	36601	46253	00477
54398	21154	97810	36764	32869
65544	34371	09591	07839	58892
08263	65952	85762	64236	39238
39817	67906	48236	16057	81812
62257	04077	79443	95203	02479
53298	90276	62545	21944	16530
62898	93582	04186	19640	87056
21387	76105	10863	97453	90581
55870	56974	37428	93507	94271
86707	12973	17169	88116	42187
85659	36081	50884	14070	74950
55189	00745	65253	11822	15804
41889	25439	88036	24034	67283
85418	68829	06652	41982	49159
16835	48653	71590	16159	14676
28195	27279	47152	35683	47280
76552	50020	24819	52984	76168
51817	36732	72484	94923	75936
90985	28868	99431	50995	20507
25234	09908	36574	72139	70185
11785	55261	59009	38714	38723
92843	72828	91341	84821	63886
18776	84303	99247	46149	03229
15815	63700	85915	19219	45943
30763	92486	54083	23631	05825
03878	07516	95715	02526	33537

Table IV. Percentage Points of the T-Distribution

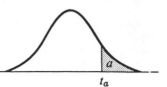

t_a

n	a = .10	a = .05	a = .025	a = .010	a = .005	df
2	3.078	6.314	12.706	31.821	63.657	1
3	1.886	2.920	4.303	6.965	9.925	2
4	1.638	2.353	3.182	4.541	5.841	3
5	1.533	2.132	2.776	3.747	4.604	4
6	1.476	2.015	2.571	3.365	4.032	5
7	1.440	1.943	2.447	3.143	3.707	6
8	1.415	1.895	2.365	2.998	3.499	7
9	1.397	1.860	2.306	2.896	3.355	8
10	1.383	1.833	2.262	2.821	3.250	9
11	1.372	1.812	2.228	2.764	3.169	10
12	1.363	1.796	2.201	2.718	3.106	11
13	1.356	1.782	2.179	2.681	3.055	12
14	1.350	1.771	2.160	2.650	3.012	13
15	1.345	1.761	2.145	2.624	2.977	14
16	1.341	1.753	2.131	2.602	2.947	15
17	1.337	1.746	2.120	2.583	2.921	16
18	1.333	1.740	2.110	2.567	2.898	17
19	1.330	1.734	2.101	2.552	2.878	18
20	1.328	1.729	2.093	2.539	2.861	19
21	1.325	1.725	2.086	2.528	2.845	20
22	1.323	1.721	2.080	2.518	2.831	21
23	1.321	1.717	2.074	2.508	2.819	22
24	1.319	1.714	2.069	2.500	2.807	23
25	1.318	1.711	2.064	2.492	2.797	24
26	1.316	1.708	2.060	2.485	2.787	25
27	1.315	1.706	2.056	2.479	2.779	26
28	1.314	1.703	2.052	2.473	2.771	27
29	1.313	1.701	2.048	2.467	2.763	28
30	1.311	1.699	2.045	2.462	2.756	29
inf.	1.282	1.645	1.960	2.326	2.576	inf.

Table V. Percentage Points of the Chi-Square Distribution

$a = .10$	$a = .05$	$a = .025$	$a = .010$	$a = .005$	df
2.70554	3.84146	5.02389	6.63490	7.87944	1
4.60517	5.99147	7.37776	9.21034	10.5966	2
6.25139	7.81473	9.34840	11.3449	12.8381	3
7.77944	9.48773	11.1433	13.2767	14.8602	4
9.23635	11.0705	12.8325	15.0863	16.7496	5
10.6446	12.5916	14.4494	16.8119	18.5476	6
12.0170	14.0671	16.0128	18.4753	20.2777	7
13.3616	15.5073	17.5346	20.0902	21.9550	8
14.6837	16.9190	19.0228	21.6660	23.5893	9
15.9871	18.3070	20.4831	23.2093	25.1882	10
17.2750	19.6751	21.9200	24.7250	26.7569	11
18.5494	21.0261	23.3367	26.2170	28.2995	12
19.8119	22.3621	24.7356	27.6883	29.8194	13
21.0642	23.6848	26.1190	29.1413	31.3193	14
22.3072	24.9958	27.4884	30.5779	32.8013	15
23.5418	26.2962	28.8454	31.9999	34.2672	16
24.7690	27.5871	30.1910	33.4087	35.7185	17
25.9894	28.8693	31.5264	34.8053	37.1564	18
27.2036	30.1435	32.8523	36.1908	38.5822	19
28.4120	31.4104	34.1696	37.5662	39.9968	20
29.6151	32.6705	35.4789	38.9321	41.4010	21
30.8133	33.9244	36.7807	40.2894	42.7956	22
32.0069	35.1725	38.0757	41.6384	44.1813	23
33.1963	36.4151	39.3641	42.9798	45.5585	24
34.3816	37.6525	40.6465	44.3141	46.9278	25
35.5631	38.8852	41.9232	45.6417	48.2899	26
36.7412	40.1133	43.1944	46.9630	49.6449	27
37.9159	41.3372	44.4607	48.2782	50.9933	28
39.0875	42.5569	45.7222	49.5879	52.3356	29
40.2560	43.7729	46.9792	50.8922	53.6720	30
51.8050	55.7585	59.3417	63.6907	66.7659	40
63.1671	67.5048	71.4202	76.1539	79.4900	50
74.3970	79.0819	83.2976	88.3794	91.9517	60
85.5271	90.5312	95.0231	100.425	104.215	70
96.5782	101.879	106.629	112.329	116.321	80
107.565	113.145	118.136	124.116	128.299	90
118.498	124.342	129.561	135.807	140.169	100

Table VI. Critical Values of the Pearson Product Moment Correlation Coefficient

	Level of significance for one-tailed test				
	0.05	0.025	0.01	0.005	0.0005
	Level of significance for two-tailed test				
$df = n - 2$	0.10	0.05	0.02	0.01	0.001
1	0.9877	0.9969	0.9995	0.9999	1.0000
2	0.9000	0.9500	0.9800	0.9900	0.9990
3	0.8054	0.8783	0.9343	0.9587	0.9912
4	0.7293	0.8114	0.8822	0.9172	0.9741
5	0.6694	0.7545	0.8329	0.8745	0.9507
6	0.6215	0.7067	0.7887	0.8343	0.9249
7	0.5822	0.6664	0.7498	0.7977	0.8982
8	0.5494	0.6319	0.7155	0.7646	0.8721
9	0.5214	0.6021	0.6851	0.7348	0.8471
10	0.4973	0.5760	0.6581	0.7079	0.8233
11	0.4762	0.5529	0.6339	0.6835	0.8010
12	0.4575	0.5324	0.6120	0.6614	0.7800
13	0.4409	0.5139	0.5923	0.6411	0.7603
14	0.4259	0.4973	0.5742	0.6226	0.7420
15	0.4124	0.4821	0.5577	0.6055	0.7246
16	0.4000	0.4683	0.5425	0.5897	0.7084
17	0.3887	0.4555	0.5285	0.5751	0.6932
18	0.3783	0.4438	0.5155	0.5614	0.6787
19	0.3687	0.4329	0.5034	0.5487	0.6652
20	0.3598	0.4227	0.4921	0.5368	0.6524
25	0.3233	0.3809	0.4451	0.4869	0.5974
30	0.2960	0.3494	0.4093	0.4487	0.5541
35	0.2746	0.3246	0.3810	0.4182	0.5189
40	0.2573	0.3044	0.3578	0.3932	0.4896
45	0.2428	0.2875	0.3384	0.3721	0.4648
50	0.2306	0.2732	0.3218	0.3541	0.4433
60	0.2108	0.2500	0.2948	0.3248	0.4078
70	0.1954	0.2319	0.2737	0.3017	0.3799
80	0.1829	0.2172	0.2565	0.2830	0.3568
90	0.1726	0.2050	0.2422	0.2673	0.3375
100	0.1638	0.1946	0.2301	0.2540	0.3211

SOURCE: Table VI is taken from Table VII of Fisher and Yates, *Statistical Tables for Biological, Agricultural, and Medical Research,* published by Longman Group Ltd., London. (previously published by Oliver & Boyd, Edinburgh), and by permission of the authors and publishers.

REFERENCES

Statistics: Concepts

Amos, J. R., Brown, F. L., and Minks, O. G. *Statistical Concepts.* New York: Harper and Row, 1965.

Bradley, J. I. and McClelland, J. N. *Basic Statistical Concepts.* Glenview, Ill.: Scott Foreman and Co., 1963.

Croxton, F. E., Cowden, D. J., and Klein, S. *Applied General Statistics.* New York: Prentice-Hall, 1967.

Freedman, D., Pisani, R., and Purves, R. *Statistics.* New York: W&W Norton & Co., 1978.

Hamburg, M. *Statistical Analysis for Decision Making.* New York: Harcourt Brace Jovanovich, 1977.

Tanur, J. M., et. al. *Statistics: A Guide to the Unknown.* New York: Holden-Day, Inc., 1972.

Wallis, W. A., and Roberts, H. V. *Statistics, A New Approach.* New York: The Free Press, 1956.

Statistics: Applications

Campbell, S. F. *Flaws and Fallacies in Statistical Thinking.* New York: Prentice Hall, 1974.

Huff, D. *How to Lie with Statistics.* New York: W&W Norton & Co., 1954.

Meek, R. L. *Figuring Out Society.* New York: Fontana, 1971.

Weeler, M. *Lies, Damn Lies and Statistics.* New York: Liveright, 1976.

Logic

Skyrms, B. *Choice and Chance.* Belmont: Dickenson, 1966.

Kahane, Howard. *Logic and Contemporary Rhetoric.* Belmont, Calif.: Wadsworth Publishing Company, Inc., 1971.

Jeffrey, R. *Formal Logic.* New York: McGraw-Hill, 1967.

Probability

Huff, D. *How to Take a Chance.* New York: W.&W. Norton & Co., 1954.

Hoel, P. G., Port, C., and Stone, C. J. *Introduction to Probability Theory.* New York: Houghton Mifflin Co., 1972.

Sample Survey Methods

Kish, L. *Survey Sampling.* New York: John Wiley and Sons, 1965.

Glaser, B. G., and Strauss, A. L. *The Discovery of Grounded Theory: Strategies for Qualitative Research.* New York: Aldine Publishing Co., 1967.

Warwick, D. P., and Linger, C. L. *The Sample Survey: Theory and Practice.* New York: McGraw-Hill, 1975.

EXERCISE ANSWERS

Chapter 1: Exercises within chapter

p. 4   Premise 1:   The birth control pill was tried on a lot of women in Puerto Rico.
       Premise 2:   None of them died.
       Conclusion: It must be safe.

p. 9   The first premise tells us that the group of unemployed women is made up *exclusively* of single women. The second premise tells us that Mary is not a member of the group of single women so she cannot be a member of the group of unemployed women.

p. 15   Defense spending includes more than just military spending, so to find out the real level of defense spending would entail finding out for example what proportion of "International affairs"; "General Science"; "Natural Resources"; and "Interest" (to name a few categories) was spending earmarked for defense purposes. Veterans' benefits and services should of course be included as a cost.

p. 18   The EQ index, as a single number, can only tell us of broad trends (as measured by changes in the EQ index). To be really meaningful, we would have to know exactly what component measures are included, such as air quality, water quality, woodland preservation, and what weight is given to each component in computing the overall index. We would also have to know how these components are measured, i.e., how is "quality" being measured, and what constitutes an "ideal environment."

p. 20   From the information given we can see at least two assumptions being made: If changing life styles, one of the three factors mentioned, does not affect the gap between male and female life-spans then the other two factors, "(women) are naturally sturdier" and "(women) take better care of themselves" must be seen as both more important and unlikely to change.

p. 21   You would need to know how the comparisons were being made—that is, whether salaries were compared by job description or overall. You would also need to know how the cost of living in Massachusetts compared with other states.

p. 22   There is no indication of what is being patented. All we know is that Gordon's Gin has a patent number and is claimed to be "smooth"—whatever that may mean.

p. 25 Inventing the light bulb, though an exemplary accomplishment, does not translate into the ability to detect a superior piano. And even if Mr. Edison was correct in choosing a Steinway lo these many years ago, it does not necessarily follow that to choose a Steinway today insures the same result.

Exercises: End of chapter

PART A

1.b. Premises: The swine-flu vaccine was tested on 10,000 people.
None of them got the swine flu.
I don't want to catch the flu.

   Conclusion: I should get vaccinated since I don't want to catch the flu.

d. Premises: On June 3, there were 24,789 people on the beach in Santa Monica.
One out of every three people had a radio on.

   Conclusion: There were 8,263 radios playing.

f. Premises: If all hockey players who got into fights were fined, there would be a significant decrease in fights on the ice.
Fighting hockey players will not be fined.

   Conclusion: No decrease can be expected.

h. Premises: In the Pennsylvania primary in 1976 Carter won more votes than any of his opponents among people from all occupations.
In the Pennsylvania primary in 1976 Carter won more votes than any of his opponents from all age groups except those over 65.
In the Pennsylvania primary in 1976 Carter won more votes than any of his opponents from all religions except Jews.
In the Pennsylvania primary in 1976 Carter won more votes than any of his opponents from both races.

   Conclusion: It was to be expected that Carter would win the election.

2.b. worthless; d. deductively valid; f. worthless; h. worthless

PART B

1.a. The premises are true as can be determined by checking the *U.S. Statistical Abstract*.

  b. We have no reason to believe that the premises are not true.

  c. The premises are true as can be checked by reading the *Boston Globe* of Feb. 1, 1980.

  d. It seems unlikely that anyone would be able to give such an accurate figure for either the number of people on the beach or the number of radios playing.

e. It is hard to believe that food prices dropped 15% in 1976. We would therefore reject the second premise.

f. Both the premises seem reasonable.

g. The premise that no woman is a good mechanic is obviously false.

h. The premises are true as can be checked by reading the *New York Times* of April 29, 1976.

2.b. Reject the conclusion. Although we have no reason to doubt that the premises are true, the conclusion does not necessarily follow. We might not get the flu, but we might die of complications (some people did) or suffer some permanent disability worse than the swine flu. Besides, who says we'll get the swine flu even if we don't get the vaccine?

d. Reject the conclusion since the premises are not plausible.

f. Reject the conclusion. Even though the premises seem reasonable, the argument is worthless.

h. Reject the conclusion. Carter's performance in one state's primary is no guide to predicting his performance in a national final election.

Chapter 2: Exercises within chapter

p. 34 $52 : 78$; $52 \div 78$; $52/78$; $2/3$; $.667$

p. 35 $24/300 = 8/100 = 8\%$

p. 35 $25¢ - 20¢ = 5¢$; $5¢ \div 20¢ = 1/4 = 25\%$

p. 36 at 1% per month 56.34¢

month	1	2	3	4	5
price (¢)	50.0	50.5	51.005	51.51505	52.550503
month	6	7	8	9	10
price (¢)	53.076008	53.606768	54.142836	54.684264	55.231107
month	11	12			
price (¢)	55.783418	56.341252			

at 1% per year 50.5¢

p. 38 $(8 + 9 + 10) \div 3 = 9$

p. 38 $PR = \dfrac{3 + 1/2\,(2)}{10} = 3 + 1/10 = 4/10 = .40 = 40\%$

Exercises: End of Chapter

PART A

1.a.
$$\begin{array}{r} \$627{,}334 \\ -573{,}832 \\ \hline 53{,}502 \end{array}$$ $\$53{,}502$

b.
$$\begin{array}{r} \$98{,}037 \\ -85{,}053 \\ \hline 12{,}984 \end{array}$$ $\$12{,}984$

3.a. $573,832/$859,449 = 66.8% (1971); $627,334/941,792 = 66.6% (1972)

  b. $85,053/$859,449 = 9.9% (1971); $98,037/$941,792 = 10.4% (1972)

5.   wages and salaries $10 - 9.3 = .7; .7/9.3 = 7.5\%$
corporate profits $17.0 - 15.3 = 1.7; 1.7/15.3 = 11.1\%$
The rate of increase for wages and salaries increased at a slower rate (7.5%) than did the rate of increase for corporate profits (11.1%).

7.   18 = 36%; 35 = 70%; 45 = 90%; 27 = 54%; 20 = 40%; 25 = 50%; 37 = 74%; 30 = 60%; 48 = 96%; 35 = 70%; 27 = 54%; 36 = 72%; 37 = 74%; 38 = 76%; 25 = 50%; 20 = 40%; 30 = 60%; 35 = 70%; 34 = 68%; 38 = 76%; 30 = 60%; 38 = 76%; 32 = 64%; 38 = 76%; 34 = 68%

9.   18: $\dfrac{0 + 1/2\,(1)}{25} = \dfrac{1/2}{25} = 2/100 = 2\%$

     30: $\dfrac{7 + 1/2\,(3)}{25} = 8.5/25 = 34/100 = 34\%$

     34: $\dfrac{11 + 1/2\,(2)}{25} = 48\%$

     38: $21/25 = 84/100 = 84\%$

PART B

1.a. The profit trend uses a poorly defined base. The profit on sales figure is misleading because we do not know return on equity. Also, profit per gallon is a flattering measure for the oil companies because they "only" make about 2¢ a gallon on average (even though this adds up to billions of dollars). Finally, there is confusion of cents with percents. (See Example 9.)

  c. Taxes represent a relatively small portion of the landlord's costs. Thus an increase in taxes of 15% does not provide a good reason for a rent increase of the same amount.

  e. These statistics tell us the percentage of men who work who hold white-collar jobs and the percentage of women who work who hold white-collar jobs. They tell us nothing of the composition of the white-collar work force by sex.

Chapter 3: Exercises within chapter

p. 54 (example 7) $7 billion

p. 55 (example 7) Procurement, Construction and other DOD

p. 56 (example 8) 6%

p. 58 (example 9)
    increase in gas & electric (1974) = 19%
    increase in gas & electric (1973, 74, 75) = 41%

Exercises: End of Chapter
PART A

3.

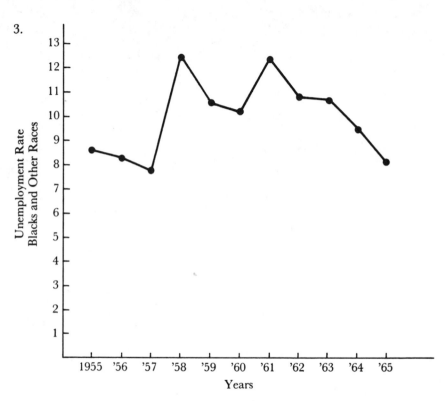

5.

Women in the Labor Force

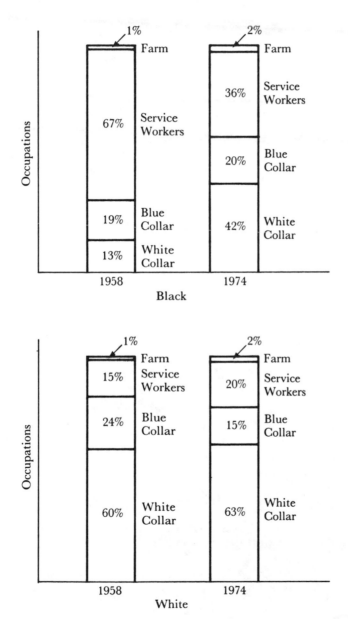

PART B

1.a. This graph emphasizes fluctuations by stretching the vertical axis. The fluctuations are also emphasized by cropping the top and bottom of the vertical scale. The main defect is that there is no indication the scale does not start at 0.

   c. The only problem with this graph is the extra space on top and the impression that it gives is that milk prices have increased *less* than they might have.

   e. In this graph the vertical axis is stretched, emphasizing the upward climb; also, the use of people figures is obviously intended to arouse emotion. Moreover, the scale begins at $8\frac{3}{4}\%$ without explaining why.

Chapter 4: Exercises within chapter

p. 78 (example 1) $\frac{\$1889.6}{\$1337.3}(100) = 141.33$

p. 79 (example 1)
   a. real output declined in 1970, 1974, and 1975
   b. decline in dollars in 1973-74 = approximately $21 billion
   c. rate of decline 1973-74: 2%

p. 81 (example 2) lowest: about 842, Dec. 1971
            highest: Dec. 1972

p. 85 a. (example 2)
   b. 0 + 1.75 + .60 + 1.50 = 3.85; 3.85 ÷ 8.00 = .48 (weight for transportation)
   c. .60 + .45 + .15 + 0 = 1.20; 1.20 ÷ 8.00 = .15 (cigarettes)
   d. .05 + .20 + .10 + .25 = .60; .60 ÷ 8.00 = .07 (newspapers)

p. 86 a. (example 2)
   b. 1967: 3.85; 1977: 0 + 3.20 + .90 + 2.75 = 6.85; 6.85 ÷ 3.85 = 1.78; 1.78 times .48 = .854
   c. 1967: 1.20; 1977: 1.30 + .70 + .30 + 0 = 2.30; 2.30 ÷ 1.20 = 1.92; 1.92 times .15 = .288
   d. 1967: .60; 1977: .15 + .45 + .20 + .50 = 1.30; 1.30 ÷ .60 = 2.17; 2.17 times .07 = .152

p. 87 5.9%

Exercises: End of chapter

PART A

1. $\frac{\$1,413.2}{\$1,214.0}(100) = 116.4$

3. $\frac{1306.6}{1.058} = 1235.0$; 1235.0 − 1171.1 = 63.9;
            63.9/1171 = .055 or $5\frac{1}{2}\%$

5. Lunch: $4.30 - 2.35 = 1.95$; $1.95 \div 2.35 = .83$
   Transportation: $6.85 - 3.85 = 3.00$; $3.00 \div 3.85 = .78$
   Cigarettes: $2.30 - 1.20 = 1.10$; $1.10 \div 1.20 = .92$
   Newspapers: $1.30 - .60 = .70$; $.70 \div .60 = 1.17$

7. Employed: b, c, d, f, a not included in the labor force
   Unemployed: e

PART B

3. Even assigning one's own weights, there are big differences in cost of living increases:

   a. it doesn't take into account a different pattern of consumption, say from steak to hamburger or from cheese to eggs;

   b. some "other" expenses rise faster than others, e.g. interest rates rise faster;

   c. personal housing costs more than doubled, but this is not reflected;

   d. medical expenses also rose far above what the average rate, because of uncovered medical bills. On a percentage basis, our personal rise was quite large, on the order of 200%.

Chapter 5: Exercises within chapter

p. 102

Class	Tally	Frequency
$80—under $100	6	6
$100—under $120	10	10
$120—under $140	7	7
$140—under $160	3	3
$160—under $180	2	2
$180—under $200	2	2

318

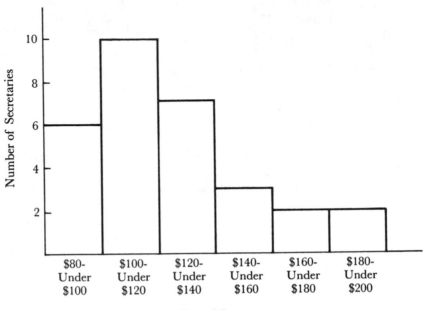

Gross Wages

p. 106
(Figure 9)
There are approximately 80 squares in the shaded area. This is about 40% of the entire area. (There are approximately 200 squares enclosed by the graph.) .12 + .17 + .11 = .40. The sum of the relative frequencies (.40) is the same as the percent of the total area (40%).

Exercises: End of chapter

PART A

1.

Class	Tally	Frequency
61–70	3	3
71–80	8	8
81–90	8	8
91–100	6	6

Relative Frequency Polygon

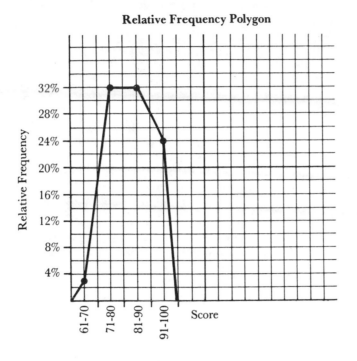

Relative Frequency Histogram

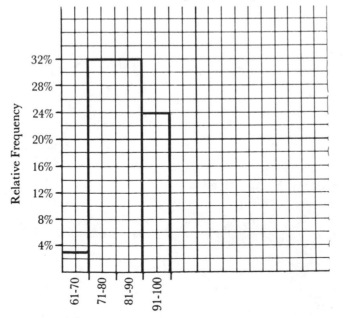

PART B

1.

Class	Grade	Class	Grade
61–70	C	61–70	D
71–80	B	71–80	C
81–90	B+	81–90	B
91–100	A	91–100	A

By assigning a grade of B to the class (71–80), 88% of the students get a grade of B or better, as 22 out of 25 scored at 71 on the test. This may be fair if the exam was extremely difficult. Alternatively, by assigning a grade of C to the class (71–80), 88% of the students get a grade of C or better. This may seem like a reasonable way to assign grades, given the scores.

2.b. By grouping the data into two classes, 21 and over and under 21, the pollster could say that while only 5% of young (read immature) people support Jones, 37% of those 21 and over do.

Chapter 6: Exercises within chapter

p. 119

Formula 1: $\bar{x} = \dfrac{\Sigma x}{n}$:

$$\frac{4+8+10+2+5+1+5+11+3+6+2+9+5+7+10+3+12+6+2+7}{20}$$

$$= 118/20 = 5.9$$

Formula 2:

$$\bar{x} = \frac{\Sigma (fx)}{n}$$

$$= \frac{(1)(1)+(3)(2)+(2)(3)+(1)(4)+(3)(5)+(2)(6)+(2)(7)+(1)(8)+(1)(9)+(2)(10)+(1)(11)+(1)(12)}{20}$$

$$= 118/20 = 5.9$$

p. 120

$50 + 51/2 = 50.5$ (average of the 50th and 51st items)

p. 121

the mode is 10; the median is 15.

p. 125

mode is 2; median is 5; mean $= (16)(2) + (11)(5) + (8)(8) + (9)(11) + (6)(14) = 334/50 = 6.7$, positively skewed; mean greater than median

Exercises: End of chapter

PART A

1. $\bar{x} = \dfrac{\Sigma(fx)}{n}$

$$= \frac{(1)(78)+(2)(89)+(3)(68)+(2)(95)+(1)(94)+(1)(62)+(1)(79)+(1)(73)+(1)(82)+(1)(74)+(1)(75)}{15}$$

$= 1199/15 = 79.27$ mean is 79.27;   median is 79; mode is 68.

3.

Class	Midpoint	Tally	Frequency
60–69	64.5	| | | |	4
70–79	74.5	┼┼┼┼	5
80–89	84.5	| | |	3
90–99	94.5	| | |	3

5. There are discrepancies in all three measures between #1 and #4. The mode changes from 68 to 74.5. The median changes from 79 to 74.5 (the midpoint of the class in which the median value is found). The mean drops from 79.27 to 77.83. All of these changes are caused by grouping the data, which results in all values in a class assuming the value of the midpoint of that class.

PART B

2.a. One problem with this quote is that if inner city is composed of people with a lot of children and people with no children, and suburbs of families with a smaller number of children the averages (means) will be the same but the distributions will be very different.

c. When talking about "permissible levels" of pollution emission levels, average values are meaningless. What you want to see is all the measured levels *especially* the extreme values.

e. If you take the mean, you should get a C, because the mean is 381/5 or 76. If you take the median, however, you shouldn't, because the median is 73. Finally, it should be noted that all grades but one are failing.

Chapter 7: Exercises within chapter

p. 133

range $25 - 4 = 21$ $(R = H - L)$

p. 138

$$\sigma = \sqrt{\frac{\Sigma f(x - \bar{x})^2}{n}}; \quad \bar{x} = \frac{(3)(3)+(5)(6)+(7)(9)+(4)(12)}{19} = \frac{150}{19}$$

$= 7.9$

x	$\bar{x}$	d	d^2	f	fd^2
3	7.9	-4.9	24.01	3	72.03
6	7.9	-1.9	3.61	5	18.05
9	7.9	1.1	1.21	7	8.47
12	7.9	4.1	16.81	4	67.24
					165.79

$$\sigma = \sqrt{\frac{165.79}{19}} = \sqrt{8.73} = 2.95$$

p. 139

Formula 3:
$$\sigma = \sqrt{\frac{\Sigma fx^2}{n} - \left(\frac{\Sigma fx}{n}\right)^2}$$

x	x^2	f	fx	fx^2
3	9	2	6	18
8	64	6	48	384
13	169	2	26	338
		10	80	740

$$\left(\frac{\Sigma fx}{n}\right)^2 = \left(\frac{80}{10}\right)^2 = 8^2 = 64$$

$$\frac{\Sigma fx^2}{n} = \frac{740}{10} = 74$$

$$\sqrt{(74) - (64)} = \sqrt{10} = 3.16 = \sigma$$

p. 139

Formula 4:
$$s = \sqrt{\frac{\Sigma f(x - \bar{x})^2}{n - 1}};$$

$$\bar{x} = \frac{48 + 102 + 44 + 216 + 96}{23} = \frac{506}{23} = 22$$

Class	f	x	$\bar{x}$	d	d^2	fd^2
10–14	4	12	22	−10	100	400
15–19	6	17	22	−5	25	150
20–24	2	22	22	0	0	0
25–29	8	27	22	5	25	200
30–34	3	32	22	10	100	300
	23					1050

$$s = \sqrt{\frac{1050}{22}} = \sqrt{47.73} = 6.91$$

p. 140

$$\frac{6.91}{22} = .31\left(\frac{s}{\bar{x}}\right)$$

Exercises: End of chapter

PART A

5. The neighborhood with a mean income of $6,000 and a standard deviation of $1,500 shows more variability of income. This is shown by the coefficient of variation, a measure of relative variability. The CV for this neighborhood is .25, while for the other neighborhood it is .20.
   $\text{CV} = s/\bar{x}$ 1. $1,500/$6,000 = .25; 2. $5,000/$25,000 = .20

PART B

3.

x	x²	f	fx	fx²
90	8100	1	90	8100
75	5625	1	75	5625
70	4900	1	70	4900
60	3600	1	60	3600
55	3025	1	55	3025
50	2500	1	50	2500
25	625	1	25	625
15	225	1	15	225
10	100	1	10	100
0	—	4	0	0
		13	450	28,700

$$\bar{x} = \frac{\Sigma fx}{n} = \frac{450}{13} = 34.6; \text{median} = 25; \text{mode} = 0$$

$$\sigma = \sqrt{\frac{\Sigma fx^2}{n} - \left(\frac{\Sigma fx}{n}\right)^2} = \sqrt{\frac{28,700}{13} - (34.6)^2}$$

$$= \sqrt{2207.69 - 1197.16} = \sqrt{1010.53} = 31.79$$

$$\text{CV} = \frac{s}{\bar{x}} = \frac{31.79}{34.6} = .92$$

The mean is a poor measure of the typical value of this distribution because there is too much variability, as the coefficient of variation (.92) shows. In addition, the extreme value of the zeroes pulls the mean way down. The mode which is 0, isn't really a typical value. The median, 25, gives a better measure than the others, since it is the middle value of the distribution. Again however, the number of zeroes (4) is the reason that the median is as low as it is. All things considered, the median is the best single measure of a typical value, but nothing here is really "typical."

5. Jane did better on her test because in her case the relative spread (as measured by the CV = .40) was less. A score ten points above the mean is more significant with a CV of .40 than with a CV of .65 (given the same units of measurement).

Chapter 8: Exercises within chapter

p. 154

between 73% and 81%: 34% of 5000 = 1750 students
between 57% and 65%: 13.5% of 5000 = 675 students

p. 158

$$z = \frac{x - \bar{x}}{s} = \frac{70 - 85}{16} = -15/16 = -.94$$

324

p. 161

$$z = \frac{x - \bar{x}}{s} = \frac{58 - 66}{11} = -8/11 = -.727 \text{ then a} = .2580$$

$$z = \frac{68 - 66}{11} = 2/11 = .182 \text{ then a} = .0793$$

.2580 + .0793 = .3373 Rounding off, we find that the relative frequency is .34 or 34%. This means we can expect scores to fall in this class (58–68) 34% of the time.

p. 162

a. $z = \dfrac{x - \bar{x}}{s} = \dfrac{35 - 39}{5} = -4/5 = -.8; \text{ a} = .2881;$

$$z = \frac{45 - 39}{5} = 6/5 = 1.2; \text{ a} = .3849$$

.2881 + .3849 = .673; 67.3% of the cars have between 35 and 45 defects.

b. 35; $z = .29$; .50 + .29 = .79; 1.00 − .79 = .21; 21% of the cars have fewer than 35 defects.

c. 45, a = .38; .50 + .38 = .88; 1.00 − .88 = .12; 12% of the cars have fewer than 45 defects.

Exercises: End of chapter

PART A

1.

3. $z = \dfrac{x - \bar{x}}{s} = \dfrac{72 - 60}{8} = 12/8 = 1.5; z = \dfrac{45 - 60}{8}$

$$= -15/8 = -1.875$$

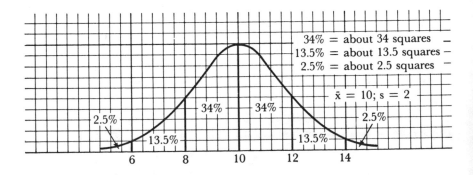

7. $z = \dfrac{x - \bar{x}}{s} = \dfrac{1.00 - .98}{.02} = .02/.02 = 1; a = .34, .50 + .34$

$= .84, 1.00 - .84 = .16$ There is only a 16% chance of buying a pound that weighs a pound or more.

9. We use the binomial distribution to solve this problem.

$p = .5; s = \sqrt{50(.5)(.5)} = 3.54.$

$\bar{x} = (50)(.5) = 25; z = 15 - 25/3.54 = -2.81; a = .4974; .50 - .4974 = .0026.$

The probability that 15 or fewer men appear in the sample is 0.26%.

PART B

1. $\bar{x} = 79; s = 10.8$ or $11, \dfrac{\Sigma fx}{n} = 79; \dfrac{\Sigma x}{n} = 80$

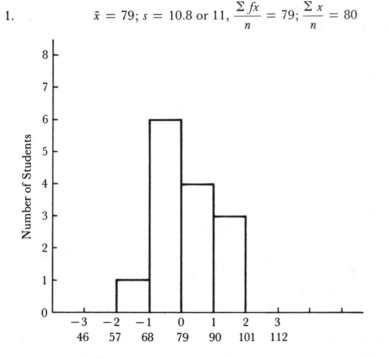

This distribution fits none of the criteria for a normal distribution: 1) bell-shaped; 2) mean = median = mode; and 3) 68% within one standard deviation, 95% within two standard deviations, and 99.7% within three standard deviations.

3.a. Not likely to be normally distributed because 52 is a small sample, but might be for some people. May also be bi-modal since food prices are lower and people eat less in the summer. Higher prices and higher consumption can be expected in the winter.

c. normal distribution.

e. The sample is too small, not random, and unpredictable.

Chapter 9: Exercises within chapter

p. 173

14th row, 2nd column, moving left: 184, 154, 238, 236, 263, 57, 203, 77, 257, 276, 298, 56, 186, 105, 271, 187, 116, 169, 70, 81

p. 176

2 age levels times 2 sexes (male and female) times 3 health levels = 12 strata

Exercises: End of chapter

PART A

1.a. $N = 550$

  b. $n = 15$

  c. You may choose to enter the table by any random means. We begin with the 10th row, 3rd column, number 29414.

  d. proceeding to the right, the 15 numbers are: 414, 532, 496, 457, 547, 136, 195, 184, 477, 398, 154, 544, 371, 263, 236.

  e. Assign each member of the freshman class a number from 1–550. Pick the people who correspond to the above numbers for the sample.

3. A housing needs survey should look at age (possibility of elderly housing); income (rent subsidies or condo conversions); and family size (large families need lots of room). Three variables that could be considered irrelevant are: race, sex, and religion.

PART B

2. There are many problems with the poll in exercise one above. First, the sample size ($n = 15$) is extremely small to be speaking for "college students." In addition, only one college is sampled and only the freshman class at that. Fifteen freshmen of indeterminate sex and socioeconomic background from any one college is not enough of a sample.

3.b. Ad executives are not likely to get a cross-section of the population by talking to their wives, friends, and neighbors, or even whoever else is available.

4.a. This question implies that welfare payments should be reduced. It is only the steps to be taken in reducing payments that is open to reply. In addition, this question uses terms like "taxpayers' money" and "welfare payments" in an emotional context.

  c. The wording invites the respondent to agree with a popular cliché.

Chapter 10: Exercises within chapter

p. 192

$25 + 60 = 85; 25/85 = .294$

p. 193

$$(.5)(.5)(.5)(.5) = .0625$$

p. 200

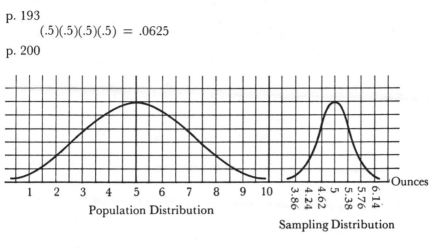

Population Distribution

Sampling Distribution

$$\mu = 5$$
$$\sigma = 1.5$$
$$n = 16$$

$$s.e. = \frac{1.5}{\sqrt{16}} = .38$$

$$z = \frac{4-5}{.38} = -2.6, z \text{ value } .495$$

$.495 + .5 = .995$ or, 99.5% of the bags will have means of 4 oz. or more.

Exercises: End of chapter

PART A

1. There is a total of 250 balls $(25 + 75 + 50 + 100 = 250)$
   a. $p(e) = 50/250 = .2$
   c. $(50/250)(25/250) = .02$

3. $p = .5; n = 25; s = \sqrt{np(1-p)} = \sqrt{25(.5)(1-.5)} = \sqrt{25(.5)(.5)}$
   $= \sqrt{6.25} = 2.5$

   In this case, 15 is one standard deviation above the mean. Therefore, assuming a normal distribution, the probability that at least 15 voters in a sample of 25 will vote for Smith is .16 (the area above one standard deviation above the mean in the sampling distribution).

5. $n = 100, p = .5, s.e. = \sqrt{p(1-p)/100} = \sqrt{(.5)(.5)/100} = .05$

   The probability is .15% or .0015 that this could happen by chance.

PART B

1.a. A blond person is more likely to have blue eyes than is a person with black hair. Therefore they do not represent simple independent events.

   c. A student who gets an A in one subject is more likely to get another A than a student who does not. Therefore getting 4 A's does not represent four independent events.

3. We are comparing the probability distribution for *one* clinic with the probability distribution for *all* clinics, thus increasing the probability of the unlikely event.

Chapter 11: Exercises within chapter

p. 216

H_0: There is no difference between the mean tread life of the sample tires and the tires in the population.

H_a: There is a difference between the mean tread life of the sample tires and the tires in the population.

p. 225

1) H_0: There is no difference between $\bar{x}$ and μ. ($\bar{x}$ equals mean tread life of the sample tires and μ equals the mean tread life of the population)

H_a: There is a difference between $\bar{x}$ and μ.

2) Significance level: $P = .05$

3) Two-tailed test—consumers would want to know if the sample tires were better or worse than the average tire in the population, at least as measured by mean tread life.

4) Z score, $z = 2$

5) Critical value: $s.e. = \sigma/\sqrt{n} = \dfrac{4,800}{\sqrt{100}} = \dfrac{4,800}{10} = 480$

$54,000 \pm 2\,(480) = 54,000 \pm 960 = 54,960$ and $53,040$

6) Conclusion: The test statistic lies in the rejection region (above the critical value). Therefore, reject H_0.

p. 231

1) H_0: There is no difference in the age distribution of the sample of widget workers and the sample of gadget workers.

H_a: There is a difference in the age distribution of the sample of widget workers and the sample of gadget workers.

2) Significance level: $P = .05$

3) Critical level: $a = .05$, degrees of freedom $= 3 - 1 = 2$; Critical value $= 5.99147$

4) Test Statistic $\chi^2 = \Sigma \dfrac{(O - E)^2}{E} = \dfrac{(46 - 52)^2}{52} + \dfrac{(43 - 39)^2}{39}$

$+ \dfrac{(11 - 9)^2}{9} = \dfrac{36}{52} + \dfrac{16}{39} + \dfrac{4}{9} = .692 + .410$

$+ .444 = 1.546$

5) Conclusion: Since the test statistic is less than the critical value, we accept the null hypothesis. That is, the difference in age between the widget and gadget workers is not significant.

Exercises: End of chapter

PART A

1.a. H_0: The landlord's effective tax rate is not different from the mean effective tax rate on all real estate.

   H_a: The landlord's effective tax rate is different from the mean effective tax rate on all real estate.

   b. Significance level: $P = .05$

   c. Two-tailed test.

   d. Z score: a $= .025, z = 2$

   e. Critical value: $\quad s.e. = \dfrac{s}{\sqrt{n}} = \dfrac{.01}{\sqrt{144}} = \dfrac{.01}{12} = .00083$

   $$.05 - 2(.00083) = .05 - .00166 = .048$$

   f. Since the test statistic is .045, and that is below the critical value (.048), we reject the null hypothesis and accept the alternative hypothesis. The landlord is getting an unfair tax advantage.

3. for #1: $s.e. = .00083$, $z = \dfrac{.05 - .045}{.00083} = \dfrac{.005}{.00083} = 6.024, P < .015$

   for #2: $\qquad P = 2(.5 - .43) = 2(.07) = .14$

   If the difference between two means is not very large, we need a relatively large sample to make sure we have a significant difference.

5. $\chi^2 = \Sigma \dfrac{(O - E)^2}{E} = \dfrac{(21 - 20)^2}{20} + \dfrac{(15 - 13.5)^2}{13.5} + \dfrac{(4 - 5)^2}{5}$

   $\qquad + \dfrac{(10 - 11.5)^2}{11.5}$

   $\qquad = 1/20 + \dfrac{2.25}{13.5} + 1/5 + \dfrac{2.25}{11.5} = .05 + .167 + .20 + .196$

   $\qquad = .613; a = .05; df = 3;$ critical value $= 7.81473$

   The test statistic (.613) is smaller than the critical value for the distribution (7.18473). The sample is therefore representative of the population.

PART B

1. Remember that we only accept or reject the hypothesis, we do not consider it proved or disproved on the basis of one experiment. Just because she found no evidence for the harmful effects of asbestos fibre in the survey does not mean that none exists.

   The hypothesis test used considered only days lost due to respiratory disease. No conclusion about the factory worker's health in general can be made from it. In addition it only compares asbestos factory workers to all factory workers, not to the general population.

Chapter 12: Exercises within chapter

p. 241

$$\bar{x} = 37.5; s = 11.0; n = 100; P = .95$$

$$s.e. = s/\sqrt{n} = 11/\sqrt{100} = 11/10 = 1.1; z = 2; 2(1.1) = 2.2$$

The estimated mean age of adults (18-65) who voted in the last presidential election in Philadelphia was 37.5 with a bound on the error of the estimate of 2.2.

p. 242

$$\bar{x} = 37.5; s = 9.65; n = 100; 95\% \text{ confidence interval}$$

$$s.e. = s/\sqrt{n} = \frac{9.65}{\sqrt{100}} = \frac{9.65}{10} = .965$$

$$z = 2; 37.5 \pm 2(.965) = 37.5 \pm 1.93 = 39.43 \text{ and } 35.57$$

The 95% confidence interval is [39.43, 35.57]

p. 243

$$\bar{x} = 4.00; s = 1.1; n = 25; 95\% \text{ confidence interval}$$

$$s.e. = s/\sqrt{n} = \frac{1.1}{\sqrt{25}} = \frac{1.1}{5} = .22$$

$$t = 2.064; 4 \pm 2.064(.22) = 4 \pm .454 = 4.454 \text{ and } 3.546$$

The 95% confidence interval is [4.454, 3.546].

Exercises: End of chapter

PART A

1.a. mean price = 72¢

  b. $\bar{x} = .72; s = .03; n = 49; z = 2; s.e. = .03/\sqrt{49} = .03/7 = .00429;$ $2(.00429) = .00858$ at a 95% probability level.

  c. $\bar{x} = .72; s = .03; n = 49;$ when we use a 90% confidence interval, $z = 1.65; s.e. = .00429; .72 \pm 1.65(.00429) = .72 \pm .0071 = .7271$ and .7129. The 90% confidence interval is [.7271, .7129].

3.a. Estimated probability of population = .89 (89%)

  b. $s.e. = \sqrt{.89(.11)/400} = \sqrt{.0979/400} = .0156$

  at 95% probability level: $z = 2 \quad 2(.0156) = .0312$, bound on the error of the estimate is 3.12%

  c. $z = 2; s.e. = .0156; p = .89$

  $.89 \pm 2(.0156) = .89 \pm .0312 = .9212$ and .8588

  The 95% confidence interval is (.9212, .8588).

5. To estimate someone's chances of being elected to office involves calculating a binomial probability. To do this we would draw a representative sample of the population likely to vote in the election and ask questions of them designed to find out their willingness to vote for this candidate. Once we had a figure, we would use the formula $s.e. = \sqrt{p(1-p)/n}$ to determine the standard error of the estimate. Assum-

ing a confidence level of 95%, we would then multiply the standard error by 2 (the *z*-score corresponding to a probability level of 95%). The binomial probability plus or minus the resulting figure would give us a 95% confidence interval, and thus a good idea of the range of the support she is likely to receive.

PART B

1.a. What seems important here is more information about the source of the estimate and size of standard error.

  c. We would want to know what kinds of jobs the under 25 people had as opposed to those over 55 because comparing estimates is misleading if the phenomena being compared are different.

Chapter 13: Exercises within chapter

p. 253

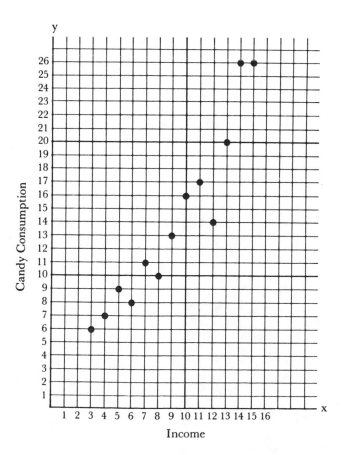

p. 255

$$b = \frac{y_2 - y_1}{x_2 - x_1} \text{; using the points } (6, 4) \text{ and } (0, 1)$$

$$b = \frac{4 - 1}{6 - 0} = 3/6 = 1/2 = .5$$

p. 258

$y = 4 + 3x$; $x = 1$; $y = 4 + 3(1) = 4 + 3 = 7$; The point $(1, 7)$ does lie on the line.

Exercises: End of chapter

PART A

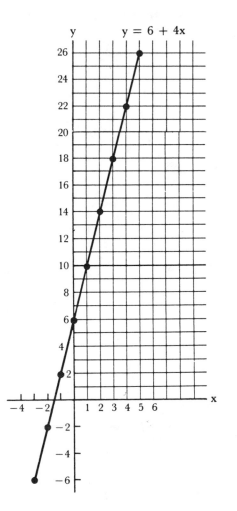

b.

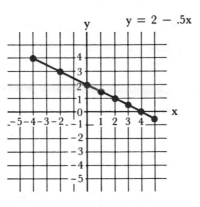

$$y = 2 - .5x$$

c.

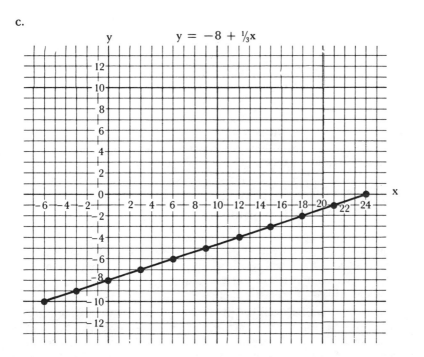

$$y = -8 + \tfrac{1}{3}x$$

3.

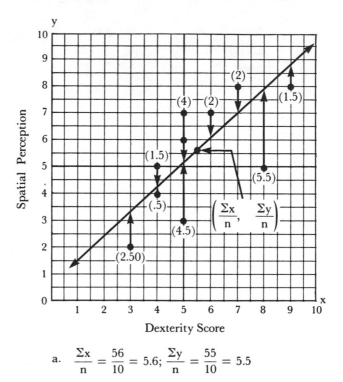

a. $\dfrac{\Sigma x}{n} = \dfrac{56}{10} = 5.6; \dfrac{\Sigma y}{n} = \dfrac{55}{10} = 5.5$

PART B

1.b. Regression analysis can be used here, with daily aflatoxin intake as the independent variable (x) being used to estimate the dependent variable (y), yearly cancer rate, but only within the extreme values of x ($x = 3.5$; $x = 222.4$), not projected beyond the given data.

Chapter 14: Exercises within chapter

p. 279

The correlation between cigarette smoking and lung cancer is positive, indicating a direct relationship between cigarette smoking and lung cancer. This means that as cigarette smoking increases, the incidence of lung cancer also increases.

The correlation between the price per pound of hamburger and demand for hamburger is negative, indicating an inverse relationship between the price of hamburger and demand for hamburger. This means that as the price of hamburger goes up the demand for hamburger goes down.

p. 283
1. **Hypothesis** H_o: There is no difference between the calculated correlation and zero.

H_a: There is a difference between the calculated correlation and zero.

2. **Significance level** = .05
3. **One tailed Test**
4. **Critical Value:** .5494
5. **Test Statistic:** .50
6. **Conclusion:** Accept the null hypothesis

p. 284

In this case, the coefficient of determination tells us that 78% of the variation in annual food expenditures can be explained by changes in annual national income.

Exercises: End of chapter

PART A

1. about .40

3.a. above 5 on verbal ability

x	y	y^2	xy	x^2
1	6	36	6	1
2	8	64	16	4
6	10	100	60	36
6	9	81	54	36
8	10	100	80	64
23	43	381	216	141

$$b = \frac{n(\Sigma\, xy) - (\Sigma\, x)(\Sigma\, y)}{n(\Sigma\, x^2) - (\Sigma\, x)^2} = \frac{(5)(216) - (23)(43)}{(5)(141) - (23)^2} = \frac{1080 - 989}{705 - 529}$$

$$= \frac{91}{176} = .52$$

$\bar{y} = 8.6 \qquad a = \bar{y} - b\bar{x}$
$\bar{x} = 4.6 \qquad \quad = 8.6 - (4.6)(.52) = 6.2$
$y = 6.2 + .52x$

$$r^2 = \frac{(6.2)(43) + (.52)(216) - 5(73.96)}{381 - 369.8} = \frac{266.6 + 112.32 - 369.8}{11.2}$$

$$= \frac{378.9 - 369.8}{11.2} = \frac{9.12}{11.2} = .81 = r^2$$

$r = \sqrt{r^2} = .90$

b. below 5 on verbal ability

x	y	y^2	xy	x^2
3	3	9	9	9
3	4	16	12	9
8	2	4	16	64
9	3	9	27	81
9	4	16	36	81
32	16	54	100	244

$$b = \frac{5(100) - (32)(16)}{5(244) - 1024} = \frac{500 - 512}{1220 - 1024} = \frac{-12}{196} = -.06$$

$\bar{y} = 3.2$

$\bar{x} = 6.4 \qquad a = 3.2 - (-.06)(.64) = 3.2 + 3.8 = 3.6$

$y = 3.6 - .06x$

$$r^2 = \frac{(3.6)(16) + (-.06)(100) - (5)(10.24)}{54 - 51.24} = \frac{57.6 - 6 - 51.24}{2.76}$$

$$= \frac{.36}{2.76} = 13$$

$$r = \sqrt{r^2} = .36$$

5. All the math for this question is shown in #3. Therefore, in a: $r = .90$, $r^2 = .81$; in b: $r = .36$, $r^2 = .13$. What r^2, the coefficient of determination does is explain the amount of variation in y (verbal ability) that is accounted for by the variation in x (manual dexterity). In a, this is 81%, in b it is only 13%.

PART B

b. What the doctor here doesn't report is that the older you are, the more likely you are to be a woman in the first place—since life expectancy is significantly longer for women than for men. The audience should have concluded that women get more chances to be sickly than men when they're old, since they're more likely to be old (and not dead). All this argument supports is that women tend to live longer than men.

c. There is a curvilinear relationship between anxiety level and test score in this case. At both low levels and high levels of anxiety, test scores suffer. But at moderate levels of anxiety, test scores are quite high. People who know that they will score low, or don't care, will have low anxiety levels. Persons who are extremely anxious will see their test scores dip. And people who are relatively calm are able to perform up to their abilities.

INDEX

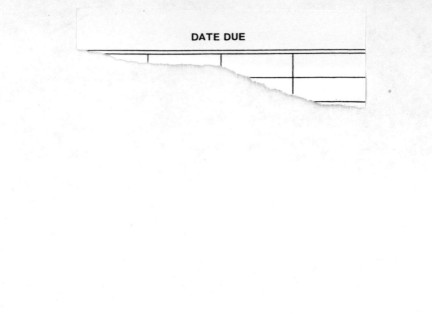

DATE DUE